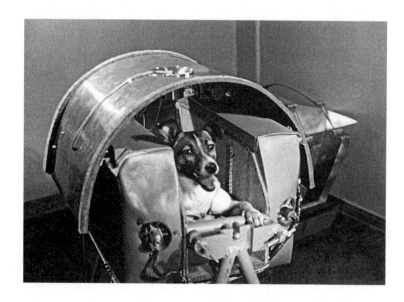

Laika's Window

The Legacy of a Soviet Space Dog

KURT CASWELL

Trinity University Press
SAN ANTONIO

for You, whose star will cross the universe
for the Dogs, their limitless kindness

Published by Trinity University Press
San Antonio, Texas 78212

Copyright © 2018 by Kurt Caswell

Jacket design by Sarah Cooper
Book design by BookMatters, Berkeley

Frontis:
Laika in a training capsule, Sputnik/Alamy Stock Photo

ISBN 978-1-59534-862-3 hardcover
ISBN 978-1-59534-863-0 ebook

Trinity University Press strives to produce its books using methods and materials in an environmentally sensitive manner. We favor working with manufacturers that practice sustainable management of all natural resources, produce paper using recycled stock, and manage forests with the best possible practices for people, biodiversity, and sustainability. The press is a member of the Green Press Initiative, a nonprofit program dedicated to supporting publishers in their efforts to reduce their impacts on endangered forests, climate change, and forest-dependent communities.

The paper used in this publication meets the minimum requirements of the American National Standard for Information Sciences—Permanence of Paper for Printed Library Materials, ANSI 39.48–1992.

Printed in Canada

CIP data on file at the Library of Congress

22 21 20 19 18 | 5 4 3 2 1

CONTENTS

O N E

◇

A Path of Burning Light

The poet's condition and the dog's is that…
they can move for a while through flame.

VICKI HEARNE
Adam's Task, 1982

April 14, 1958, must have been a particularly clear night, one
of those nights you remember for the great wash of the cosmos overhead, and for the black blackness of interstellar space
punctuated by stars uncountable. It must have been so, for
along the eastern seaboard of the United States, and out over
the Caribbean, out to the east of the islands of Saint Thomas,
Antigua, Martinique, Saint Lucia, Barbados, Trinidad, on to
British Guiana (now Guyana), and over a host of ships in the
Atlantic positioned between 10 and 20 degrees north latitude,
people gazing skyward saw a path of burning light. Some reported that they had seen a comet. Others said it was a meteor,
the fiery path of a meteoroid burning in the Earth's atmosphere, what we like to call a shooting star. And still others
reported they had seen a UFO.

But what those people witnessed was not a comet or a shooting star, and it was not a UFO. It was a phenomenon that had occurred only one other time in Earth's history: the reentry into Earth's atmosphere of an artificial satellite, in this case the Soviet Union's second satellite, *Sputnik II*. On board was a small, white mongrel dog from the streets of Moscow. Her name was Laika. One of dozens of trained space dogs in the USSR's new space program, she was the first living being to orbit the Earth. And as the satellite *Sputnik II* came down, Laika was already dead. Her body, in a state of decay from the cooling and warming inside her capsule as it passed in and out of the influence of the sun, had been in orbit around the Earth for the previous five months, traveling at a speed of 17,500 miles per hour and making 2,570 revolutions. And as *Sputnik II*'s orbit decayed and it fell into the friction-wall of the atmosphere, it burned, and the world's first space traveler, Laika, burned with it.

In the twenty-first century, such satellite reentry events have become almost routine. The National Aeronautics and Space Administration (NASA) reports that there are more than twenty-one thousand artificial debris objects in Earth's orbit larger than ten centimeters, and half a million more measuring between one and ten centimeters. Even smaller objects, smaller than one centimeter, number over one hundred million. According to NASA, such debris includes derelict spacecraft, the upper stages of launch vehicles, payload carriers, debris intentionally released during mission operations, debris created by spacecraft or upper-stage launch vehicle explosions and collisions, solid rocket motor effluent, and flecks of paint from various spacecraft or spacecraft parts released by thermal stress and small-particle impacts. The space around our planet, like the planet itself, is awash in our garbage. This space junk burns up in the atmosphere all the time, some of it visible, some of it not, sometimes as planned by the object's creators (Russia's space station *Mir* in 2001, for example) and sometimes in its own time or due to a loss of ground control (China's *Tiangong-1* [Heavenly Palace] space station, which came down in 2018). The United States Space Command tracks scheduled and nonscheduled reentries, and if you are in the right place at the right time you might see one, if you can be bothered to look up.

But in 1958 when *Sputnik II* came down, such an artificial satellite reentry event had happened only once before, and that was when *Sputnik I,* the first artificial satellite ever, burned up in the atmosphere on January 4 of that year. The presence of something in orbit around the Earth that human beings had made was still in the realm of the fantastic. It occupied the fictional worlds of writers like Jules Verne and Arthur C. Clarke and the imaginations of scientists like Russia's Konstantin Tsiolkovsky and America's Robert Goddard. And so fantastic was the idea of one of these satellites coming down so that you could see it come apart across the sky, Italian-born American astronomer L. G. Jacchia made a journey to various key points along the path of *Sputnik II* to interview people and collect data. He filed Special Report 15, "The Descent of Satellite 1957 Beta One," with the Smithsonian Astrophysical Observatory.

The story Jacchia tells is that in the early hours of April 14, *Sputnik II* fell out of orbit following a path from New York

City to the mouth of the Amazon River. Observers in the eastern United States reported seeing a long, streaking tail pass
overhead, not yet very bright, and followed by sparks, burst-
ing and shuttering away. The satellite passed over Long Island
and then went unreported for some five minutes before being
picked up again at about 23 degrees north latitude. By the time
it reached Antigua in the Caribbean, it had fallen to about 50
miles above sea level (down from its orbit at about 131 miles
perigee and 1,030 miles apogee), and it had become self-lu-
minous, a fireball rivaling the brightness of the moon. It fell,
burning, pulling a long tail behind it, sparks flying off into
the blackness. The head of that streak of light, the satellite
itself, still carrying the body of Laika, glowed white with tinges
of blues and greens through to the tail's yellowish fire that
degraded to oranges and reds out to the end. Pieces of the satellite broke off and burned alongside the main body, before
dimming and dropping away. Observers on one of the ships
at sea, the *Regent Springbok*, reported that the satellite looked
like the tail of a peacock, "each particle glowing through the
spectrum from white to a deep blue in magnificent display."
When the satellite reached about 11 degrees north latitude,
east of Trinidad and Tobago, and had fallen to about 35 miles
altitude, it exploded in a fiery burst, like great fireworks lighting up the dark. In the moments after that burst, an eerie pale
light was reported, illuminating the decks of ships at sea and
the sea around them. What was left of *Sputnik II* and the first
space voyager, Laika, traveled on, falling and burning in its
arc across the Atlantic and over Suriname and French Guiana, then onto the eastern shoulder of Brazil. As the satellite

burned, the dog burned with it. The dry calcium phosphates of Laika's bones, the salts and minerals and the carbons of her body—the very building blocks of life—dissipated in the upper atmosphere to drift on stratospheric winds. Eventually some of the matter that had once been Laika, now vaporized and elemental, rained down onto the Earth, where her life began. And somewhere out there headed toward the place where one of the great rivers of the world, the Amazon, meets the sea, *Sputnik II* vanished completely, still traveling fast above the horizon line. It burned out and was reconsumed by the great black nothing of the cosmic dark. The entire event unfolded in about ten minutes.

❏

The year *Sputnik II* came down, the Soviet Union and the United States were locked in a bitter stalemate known to us as the Cold War, a stalemate pitting Soviet communism and American capitalism against each other, which dominated political, economic, cultural, and even religious patterns between the two countries and across our world. The year *Sputnik II* came down, the human population of our planet was just 2.9 billion. And because there were relatively few of us, there was a whole lot more of everything else. More rhinos, more trees, and more clean water and air. The world was recovering from its second world war, and a growing middle class in countries like the United States and the United Kingdom was moving us all forward into a global consumer economy the likes of which had never been known. It didn't have to go that way, but it did. Not in two hundred thousand years of the human story had so many people had so much. Not ever. And probably not ever again.

In 1958 the British Overseas Airways Corporation (now defunct) established the first transatlantic jet service, flying between New York and London. Such flights are routine in the twenty-first century, so much so that we call this astonishing feat of engineering and science "crossing the pond." The trip that took Columbus and his three ships over two months to achieve was now possible in about eight hours, and available to almost any middle-class citizen of any country in the world. And we have gone faster still. Astronauts on a space shuttle flight out of Cape Canaveral, Florida, easily crossed the Atlantic in about nine minutes. Also in 1958, the European Economic Community was established, the precursor of the European Union. The New York Yankees defeated the Milwaukee Braves to win the World Series. Elvis was inducted into the US Army. And in the US, a stamp cost three cents, and a gallon of gas cost a quarter.

In 1958 Dwight Eisenhower was president of the United States, and Nikita Khrushchev was named the new Soviet premier. This was the year Russian novelist Boris Pasternak won the Nobel Prize in Literature but declined to accept the medal at the ceremony in Sweden under pressure from his nation's government. The Soviet Union had banned publication of Pasternak's novel *Doctor Zhivago* two years earlier. Meanwhile the CIA was running a propaganda campaign and played a central role in the novel's publication in Russian (it was first published in Italian translation) to push its perceived anti-Soviet threads into the ring of global politics. This was also the year American pianist Van Cliburn won the first Tchaikovsky International Piano Competition, held in Moscow. His final performance of Tchaikovsky's Piano Concerto no. 1 and Rachmaninoff's Piano Concerto no. 3 brought the primarily Soviet audience to its feet, and they stood applauding for eight minutes. The competition was intended to bring Soviet cultural superiority to the world's stage, but instead that spotlight was stolen by an American, a Texan no less, whom *Time* later touted on its May 19, 1958, cover as "The Texan Who Conquered Russia." In this now famous story, the judges were compelled to ask Premier Khrushchev for permission to award first prize to an American. "Is he the best?" Khrushchev is said to have asked. The judges consented that he was. "Then give him the prize," Khrushchev said.

The prize was given, and in fact it was given on the same ordinary Monday that *Sputnik II* came down. Clearly there was a lot going on that night, and clearly there was a lot going on in 1958. But *Sputnik II*, together with *Sputnik I*, towers above most everything else during that year, or during that decade, and arguably during the several decades before or after. That satellite with the little dog inside—its launch, its orbit, and its burning in the atmosphere—is one of the events that has redefined *Homo sapiens* as a species. That event signals a singular moment of self-determination when we first left our planet home to venture into the cosmos, crossing into what Aldous Huxley, and Shakespeare before him, called a "brave new world." *Sputnik I* and *II* are the first steps in becoming an interplanetary species, in becoming Earth independent. In his book *Soviet Space Exploration*, William Shelton writes of Russian novelist

Vladimir Orlov's determination that artificial satellites and spacecraft are artificial planets because they are populated by all manner of creatures from Earth, including humans. From the time of the first two *Sputniks*, humans have maintained a continuous presence in space, living and working in low-Earth orbit on various space stations, walking on the moon, and operating robotic rovers, telescopes, satellites, and probes as far out as Pluto, and even beyond. From these artificial planets, then, we will continue on and colonize a natural planet, probably Mars. Indeed, private corporations and space agencies in a number of countries have their sights set on a crewed mission to Mars. And it may happen sooner than you think. "There was a time," writes Orlov, "when terrestrial life stepped over the threshold of the ocean and conquered the land; now it has stepped off the Earth to conquer the abyss of the cosmos."

What the Soviets did first, and what other nations would do after, resides at the zenith of our curiosity, because to be human is to be curious, to be an explorer. We cannot help but look outward to the next horizon, to the far-off and beyond, to the distant and the fantastic. We cannot help but dream. It is what we are, and what we do, and it tumbles from the beginnings of our biological evolution. In his essay "The First Earth Satellite," Sergei Khrushchev (son of then-Soviet premier Nikita Khrushchev and a rocket engineer in his own right) calls this drive to explore "the naïve confidence that [we] are equal to anything." Human beings embarked upon space exploration, Khrushchev writes, to prove to ourselves "that [we] can touch the stars with [our] hand, that the Moon is only a stopping station, the first step, and that the next step is Mars, and after this anywhere. For me," he writes, the time of *Sputnik* "was the best and the brightest time of my life." The Space Race, during those early years, and even now, writes Khrushchev, "was a race not to the death, but to immortality."

◻

Until Laika's flight, scientists did not know what would happen to a living organism outside the protection of Earth's atmosphere where there is no oxygen to breathe, weakened gravity, and increased radiation. What would be the effects of solar radiation (from our sun) and cosmic radiation (from outside our solar system) on a living organism in orbit? The Van Allen belts had yet to be discovered, and their role, along with Earth's magnetic field, in shielding the Earth and its atmosphere from radiation was not known. How long, scientists wanted to know, could a living organism survive in orbit? Five minutes? Three days? One year? No one knew. And what would be the effects of microgravity, or weightlessness, on a living organism? Could the body's organ systems function in microgravity? Or would the whole thing just shut down? No one knew. Oleg Gazenko, a physician who helped select and train Laika, notes in an interview for the BBC documentary *Space Dogs* that "it was absolutely essential to have an answer to the question, was weightlessness really an insurmountable barrier to the chances of a human surviving any length of time in the conditions of space travel?" And even if a human being could manage increased radiation and decreased gravity, could he survive the flight into space, the g-force of an accelerating rocket, and the violent vibration of that wild ride?

No one knew. So before sending a human being into orbit, we sent Laika, a little white dog from the streets of Moscow, who would test these unknowns for us. "Quite simply," writes Olesya Turkina in her book *Soviet Space Dogs*, "without the first dog in space there would be no human spaceflight."

Laika rode into orbit on November 3, 1957. The USSR reported that she survived for about a week, returned a stream of valuable data that would help make human spaceflight possible, and then died a painless death as her oxygen ran out. Soviet chief designer Sergei Korolev announced in a statement that "the data gathered on cosmic rays during the flight of *Sputnik II* [was] of great value" and that "the study of biological phenomena made during the spaceflight of a living organism—something done for the first time in *Sputnik II*—[was] of tremendous interest.... The time will come when a spacecraft carrying human beings will leave earth and set out on a voyage to distant planets—to remote worlds. The way to the stars is open."

○

When Laika came to the kennels in Korolev's space dog program, the team of engineers and scientists first knew her as Kudryavka, Russian for Curly, or Little Curly. A few sources report a nickname, Zhuchka, or Little Bug. After she went into space, a few Soviet sources referred to her as Limonchik, or Little Lemon, but that name fell away rather quickly. Then, capitalizing on the name of the satellite itself, the American press came to call her Mutnik, a mongrel satellite. But the name by which she is best known, her true and proper name, is Laika, the first living being to orbit the Earth, and the first to die out there too. *Laika* is a noun derived from the Russian verb *layat*, which means to bark. So in Russian, Laika means Barker, or Little Barker. The word *laika* also refers to a breed of dog, a medium-sized hunting dog of northern Siberia, of which there are several types. Laika herself may have come from laika stock, but it would be incorrect to call her a laika. She was a mixed breed, a mongrel, a throwaway of unknown origin, living off the scraps and refuse of Muscovites.

That Laika barked, or was a barker, is not in question, but what did her barking mean to the Soviet team that trained and worked with her? Is her name an expression of their annoyance or impatience with her and her barking? Or is it a celebration of her personality, her character, an identifier as a vocal dog, a dog that speaks, a dog that communicates because she is in tune with her surroundings and the dogs and people who interact with her? Perhaps the team came to regard Laika's barking as a quality that distinguished her from the other space dogs in the kennels, a characteristic that made her stand out.

At the time of her flight, Laika weighed thirteen pounds and was about two years old, the ideal size and age for a space dog. Her fur was mostly white, with a darker brown covering her face, and a circle of white running around her black nose and leading up between her eyes to the crown of her head. Her ears stood straight up, like a lot of laika breed dogs, but then bent over at the tips, giving her a friendly look. In video footage of Laika, her bent ears bounce about as she sits or stands, panting, giving her an air of nervous ease. In her eyes, though, is an attendant intelligence, a quiet confidence, the look of a dog that is deeply attuned to the people around her, to what they are doing, and what that means to what she is doing.

A memorial sculpture in Moscow depicts Laika with her tail curled up over her back, another common trait among laika breeds, and this curly tail might be the reason for initially calling her Little Curly. But in most of the photographs of Laika I have seen, her tail hangs straight down like a German shepherd. In these photographs, though, Laika is wearing her flight suit and harness with its waste collection bag. In *Inside of a Dog*, Alexandra Horowitz writes that dressing a dog in clothing (a sweater or a raincoat, for example) reproduces the feeling of another dog standing over it, pressing down on it in a show of dominance. Perhaps Laika's posture in these photographs, then—standing frozen in place, her tail dropped straight down—registers discomfort and submission. Presumably when she was free and running about, or out on a walk with her handlers, her tail rolled up into that nice little curl.

Laika's story comes into the historical and cultural record when she becomes a space dog in training. As a stray living on the streets of Moscow, she was acquired by physician Vladimir Yazdovsky's team, the man Korolev had put in charge of directing biomedical operations in the emerging space program, which included training and caring for the space dogs. If Laika had been on some other Moscow street that day, or hidden away in a quiet alley, or had she simply lived in some other part of the city, she would have been just another stray, faceless, nameless, living and dying in her time. Her story would have been so unremarkable as to be lost to us, as are the stories of most of the living creatures, humans included, across the ages of the Earth.

While we cannot know much about Laika's life before she came to Korolev's kennels, we can imagine it. In his graphic novel *Laika*, Nick Abadzis does imagine it, and it allows us a window on the early life of the dog that took us to the stars.

In Abadzis's book, Laika is born into a litter of seven puppies in the house of a government official. The housemaid is ordered to get rid of the pups. One of the female pups becomes her darling, and the housemaid works hard to find her a proper home. A family adopts the pup for their young son. The boy resents the little dog for the way she demands his attention and time, and one night he tosses her into the Moscow River. She swims to the bank, a castoff, and takes to the streets. She befriends another stray dog. When a team of city dogcatchers captures her, they kill her companion in the process. A sympathetic dogcatcher peers in at the pup through the door of her cage and remarks that she "looks like [she's] been through the wars." The animal shelter is full, and the pup will have to be euthanized, but then the dogcatcher remembers Yazdovsky, "the air force chap," who is looking for small stray dogs for some secret government program. That secret government program is, of course, the space program, the program developing missiles and rockets, and training dogs to ride those rockets into space. And so the pup becomes a space dog. The pup becomes Laika.

In Abadzis's book, the housemaid keeps Laika in her heart and thoughts, even after she is gone, and Laika too bears a memory of the housemaid. Neither knows where the other is, but their bond remains strong, for it is the bond that tethers those who go with those who stay behind. "Never look back," Abadzis writes as Laika roams the Moscow streets. "Although those you leave behind will still think of you. We do still think of you."

I have to wonder, when Laika was a street dog in Moscow, did someone feed her, help her, give her a warm place to sleep on a cold winter night? Did a kind restaurateur give her food scraps when she appeared at the alley door? Was that someone a man, a woman, or several men or women? Was Laika befriended by a child? Did these people I am imagining mark that day when Laika, who was always there, was suddenly gone? And living through those next decades, and to the end of their lives, did they know (how could they know?) that the stray dog from the streets they had become so fond of was the same dog they read about in the newspapers and heard about on the radio, the now world-famous dog orbiting the Earth?

Despite the possibility of such a guardian angel attending to Laika on the streets, life inside Korolev's kennels would have been a marked improvement. Of course, she had to endure space dog training, and eventually the job she was trained to do, but in exchange, she lived in a warm, dry enclosure with wooden floors, with clean bedding of wood shavings or straw. She had the company and social interaction of other space dogs in training, most with similar backgrounds and stories. Her handlers took her out for a walk at least twice a day. And she was watered and fed regularly, a diet that included meat, bone broth, vegetables, fish oil, and milk. Space dogs that were about to fly in a rocket were offered an even better meal that sometimes included a good Russian sausage. This practice was carried forward to cosmonauts and astronauts, who gather even today for a preflight meal and may request anything they desire from the kitchen.

◻

When *Sputnik II* went up, the Soviet Union had already beaten the United States into orbit with *Sputnik I*, the world's first artificial satellite, establishing the Soviets' technological superiority, and so their supposed superiority as a people and a nation. "Success in space," writes James Oberg in *Red Star Rising*, "implied superiority on earth." This feat is even more astonishing considering the ruinous state of the USSR at the end of World War II. "No other nation in the world was as devastated and crippled by the war," writes Fordham University space historian Asif Siddiqi in *Challenge to Apollo*. The war killed some twenty-seven million Russian soldiers and citizens and destroyed some 1,700 cities along with half the nation's housing. The agricultural system "was close to famine proportions." Astonishing that out of this kind of ruin, it took the USSR just over a decade to put *Sputnik I* into orbit. But however astonishing the sight of that satellite in orbit, it was tempered in those days by an equally awesome terror rising out of this new technology, a technology that enabled one nation to destroy another with the touch of a button. As the world's first intercontinental ballistic missile (ICBM), the rocket that carried *Sputnik* into orbit could also carry a nuclear warhead to nearly any place on Earth.

The Soviets called their satellite *Prosteyshyy Sputnik-1* (Simplest Satellite-1) or *PS-1*. The world came to know it as *Sputnik*, the Russian word for "satellite," or more precisely "traveling companion" or "following companion." When an object achieves orbit, it becomes a satellite, for the word "satellite" defines any body in orbit around another. In the Middle Ages "satellite" was used for a person who follows another person superior in rank or status. Astronomers took control of the word in the early days of astronomy, using it to describe natural bodies in orbit, like Earth's moon, which is a satellite. The successful launch of *Sputnik I* demanded a distinction between artificial satellites and natural satellites. Soon the word "satellite" fell out of popular use for natural bodies and came to mean "artificial satellite," almost exclusively. Now the moon is just a moon.

Satellites do not generally follow a circular path around the Earth but rather elliptical paths of varying degrees, just as the Earth and the other planets in our solar system follow elliptical paths around our sun. Perigee is the distance of a satellite at its closest point to the Earth, and apogee is the distance at its farthest point. A satellite's orbit can change over time too, especially if its perigee is low enough that it drags in the upper atmosphere. Eventually, such atmospheric resistance will bring the satellite down. The International Space Station (ISS), in low-Earth orbit, has a perigee of about 254 miles and an apogee of 258 miles, not much of an elliptical path at all. It is moving very fast, at about 17,500mph, and makes one lap around the Earth every ninety minutes. For comparison, some satellites are in highly elliptical orbits (HEO), somewhere around 600 miles at perigee and over 22,000 miles at apogee. Such satellites spend most of their time moving out to and back from apogee, giving them a longer dwell time over a specific region of the Earth. Nations in extreme northern and southern latitudes often place communications (and other) satellites in HEO. The technique was pioneered by Russian engineers out of necessity, as theirs is a nation with a great deal of landmass in the far north. Weather satellites are placed into geostationary orbits (GEO), which means they orbit at a bit more than 22,000 miles from the Earth. At this distance the satellite's speed matches the rotation of the Earth, and so, despite the fact that it is moving really fast, from the ground it doesn't look to be moving at all. Still farther out, the moon has a perigee of 226,000 miles and an apogee of 252,000 miles. In its orbit around the sun, the Earth with all its satellites in tow, natural and artificial, has a perihelion (*helios* from the Greek for "sun") of 91.4 million miles and an aphelion of 94.5 million miles.

Sputnik I was a small satellite, a polished metal sphere only twenty-three inches in diameter (a bit more than double the size of a basketball), with four antennas positioned equidistant from each other at their base, canted and pointed back, all in the same direction. Launched on October 4, 1957, *Sputnik I* broadcast a beeping signal at 20.005 and 40.002 megahertz, easily picked up by government and amateur radio operators worldwide. After twenty-one days, its batteries ran down, and *Sputnik I* went silent, but it continued to orbit for two more months, most of that time concurrently with *Sputnik II*.

Premier Khrushchev, on his way home from vacation on

the Black Sea—a region of splendid beaches and picturesque mountains still coveted as a vacation spot by Russia's elite, including President Vladimir Putin—stopped in Kiev to take in news of the launch. He was impressed by the technological feat of *Sputnik I,* but then there were many technological feats. He did not understand the satellite's importance until the next day when the world responded. In his essential book *Sputnik: The Shock of the Century,* Paul Dickson reports that newspapers, TV news, scientists, politicians, practically everybody congratulated the Soviet Union, heralding the event as one of the world's greatest scientific achievements. It was also lauded as a propaganda stunt to advance the Soviet Union's position in the world, one that worked most effectively. All over the world people turned out by the millions along the path of the satellite to watch it pass overhead. Millions more picked up its now iconic signal—"beep, beep, beep, beep"—on ham radios or listened to its signal broadcast on a local radio station. Dickson reports that British physicist Sir Bernard Lovell called *Sputnik I* "about the biggest thing that has happened in scientific history." And the British science fiction writer Sir Arthur C. Clarke, who also holds his own as a science writer, called the satellite "one of the greatest scientific advances in world history" and announced that it would have "colossal repercussions."

As a young man, Gil Moore, an American rocket engineer who worked on the Viking and Aerobee projects that were later modified and became the first two stages of the navy's Vanguard launch vehicle, saw *Sputnik I* pass overhead in the night sky. Now in his early nineties and wearing a black patch over his left eye, he told me the story sitting at his kitchen table in his Colorado home:

> All the people who said they saw *Sputnik I* are incorrect. The
> final rocket stage was a great big long cylinder, and it was also
> in orbit. You could see it tumbling end over end as it reflected
> light from the sun. People who saw that thought they saw *Sputnik*
> itself, but they didn't. *Sputnik* was too small to see with the eye.
> It was the final stage of the rocket that people saw. We were
> standing outside our home near Las Cruces [New Mexico],
> my wife and I. And when that sucker came over, my mind
> was blown. Because here was this enormous flashing coming
> overhead, and there was no sound. This thing was totally silent.
> I was not listening to the beep, beep, beep. To me, *Sputnik*
> was eerily silent. I thought, oh man, that is an ICBM, the basic
> launch system. And I thought, those suckers can drop bombs
> right on our heads. So yeah, yeah we saw it come over. And it was
> really quite a sight in the sky.

Norm Augustine, former chairman and CEO of Lockheed Martin, told me he was in his first week of graduate school in aeronautical engineering when someone walked up to him and announced that the Soviets had just launched a sputnik. His response was, What's a sputnik? Then, Augustine said, "upon learning the answer, coming as it did at the height of the Cold War, it was like a blow to the gut."

Beyond its military applications, the push to get a satellite into orbit was a science project, part of a global scientific cooperative called the International Geophysical Year (IGY), which ran from July 1, 1957, to December 31, 1958. Scientific collaboration between the USSR and the United States was impossible under Soviet premier Joseph Stalin, but his death in 1953 changed that. Sixty-seven countries took part in experiments and advancements in eleven areas of Earth science, including cosmic rays, geomagnetism, gravity, ionospheric physics, meteorology, and solar activity. The timing for such a global collaboration was ideal, because it encompassed the peak of Solar Cycle 19, a roughly eleven-year cycle of changes in the sun's activity. Solar cycles are numbered starting in 1755 but have been reconstructed back to the beginning of the Holocene, which marks the end of the most recent ice age. As much as the IGY was about collaboration, it was also highly competitive, especially between the USSR and the US. Both countries announced that they would put an artificial satellite into Earth orbit during the IGY, but no one thought the Soviets could do it, and especially not do it first. No one had any idea the Soviets were so far along in their rocket and missile program, Moore told me, because "at the time, the Soviets were incredibly secretive about their work. We knew nothing about it in those days." The perception was that the US would lead the way into space, because the Soviets were still fastened to the previous century, just a nation of poor farmers and laborers suffering through relentlessly cold winters. When "the USSR announced that they were going to join the IGY," Moore said, "scientists and engineers in the US mocked them. We thought they didn't know anything. And then they put up *Sputnik*."

Sputnik I left the US scrambling to put up its first satellite, *Explorer I*, which would not be achieved for nearly three more months. *Explorer I* was a much smaller satellite, weighing just over thirty pounds, while *Sputnik I* weighed in at 184 pounds. *Sputnik II* topped out at over 1,100 pounds. For comparison, the GOES-R weather satellite that I saw under construction in a clean room during my visit to Lockheed Martin's Denver campus weighs 11,500 pounds. It launched from Cape Canaveral, Florida, on a United Launch Alliance Atlas V rocket in 2016. In the early days of the Space Race, however, the US did not have a rocket with the kind of power necessary to lift an object as heavy as *Sputnik II* into orbit, and Khrushchev knew it. *Vanguard I*, the US's second satellite, went up on March 17, 1958, and it was even smaller than *Explorer I*. While it bears the distinction of being the first satellite powered by solar energy, it weighed a mere 3.2 pounds. Khrushchev taunted the US, calling *Vanguard I* a grapefruit. Even so, while the first two sputniks came down within a couple of months, *Vanguard I* is still up there, and it will likely remain in orbit for another couple hundred years.

Some historians have written that President Eisenhower was much less concerned with being first into space than he was in establishing space as open and free to all nations. Such an international agreement, Eisenhower knew, would allow nations to fly satellites over other nations, which would be very useful in spying. Along with other senior officials, Eisenhower had the advantage of reports from a new high-altitude spy plane, the U-2, which revealed that the Soviets were really not technologically or militarily ahead of the US. There was really no good reason to panic; still, his plan was a calculated risk. If the US was first into space and its satellite flew over the USSR, he reasoned, the Soviets might claim the satellite had violated their sovereignty and would then urge a referendum to partition not only the skies over nations, but outer space too. However, if the Soviets were first to fly their satellite over the United States, and the US did not protest, they would then have set a precedent enabling the US to fly satellites over the Soviet Union. What Eisenhower could not have predicted was the reaction of the American people to being beaten into space by the Soviets. In the US, the initial excitement surrounding the technological achievement of *Sputnik I* was supplanted by fear and anger, which was bad for the American psyche, bad for American security, and bad for Eisenhower's presidency.

The US had, in fact, two leading rocket programs in operation at the time: the navy's Project Vanguard (which Gil Moore worked on) and the army's Redstone missile program, directed by the German engineer Werner von Braun. Eisenhower suspected that putting a satellite into orbit on a military-purposed Redstone missile would incite fear in the Soviet Union, so he endorsed Vanguard's science mission and told von Braun and his team to stand down. But Vanguard struggled. Attempt after attempt ended in failure. Von Braun's team launched a rocket weeks in advance of *Sputnik I* that could have entered orbit, but they pulled it back to comply with the president's directive. Then *Sputnik I* went up. "For God's sake, turn us loose," von Braun said. Eisenhower finally did, and *Explorer I* launched into orbit on January 31, 1958, on von Braun's Redstone rocket, which put the US back on track in what became known as the Space Race.

Before the successful launch of *Explorer I*, though, Khrushchev planned to stun the world again with Soviet power and ingenuity. He ordered the launch of a second satellite, *Sputnik II*, to coincide with the fortieth anniversary of the Bolshevik Revolution. The satellite was also a science laboratory, as Laika, along with *Sputnik II*'s instruments, would return data to the USSR essential to understanding the conditions in Earth orbit, data crucial to finding out if and how a human being could survive in space. The Soviet news agency TASS issued this statement in a press release: "Artificial earth satellites will pave the way to interplanetary travel, and apparently our contemporaries will witness how the freed and conscientious labor of the people of the new socialist society makes the most daring dreams of mankind a reality."

◻

During her flight and for decades after, Laika was one of the most famous dogs in the world. She appears on most every list of famous dogs, along with Lassie, Hachiko, and Rin Tin Tin; and she is listed among *Time* magazine's fifteen "most influential animals that ever lived," joining the company of Alexander the Great's war horse, Bucephalus; Dian Fossey's favored mountain gorilla, Digit; and the world's first successfully cloned adult mammal, Dolly the sheep. The Soviet Union issued a postage stamp in Laika's honor, and so did Albania, Benin, North Korea, the Emirate of Sharjah (part of the United Arab Emirates), East Germany (now the reunited nation of Germany), Guyana, Hungary, Mongolia, Nicaragua, Poland, and Romania. Laika-brand cigarettes were hugely popular in the Soviet Union and in other countries. Her image was featured on cigarette cases, cigar bands, matchboxes, postcards, posters, in newspaper and magazine drawings and cartoons, on boxes of chocolates and chocolate wrapping papers, lapel pins and badges, handkerchiefs, confectionery tins, playing cards, commemorative plates, desktop sculptures, and porcelain figurines. In Japan, Laika's image was featured on a child's tin watering can, a spinning top, and a bucket. In the US she was featured on a child's piggy bank, or "Sputnik Bank," and on a child's toy plastic helmet with two metal spring antennas, the "Wee Beep Sputnik" helmet. In West Germany a child's mechanical toy featured Laika in a sputnik orbiting the Earth. And in Mexico, a tin serving tray pictured Betty Boop walking Laika on a leash across the surface of an alien world, possibly the moon. Laika has been the subject of poems, children's books, at least one graphic novel, a few books of nonfiction, songs, and music videos. Years later, circulating on the internet, is the curious theory that Scooby-Doo is an escaped Soviet space dog, perhaps in Laika's image, and you can see such a space dog running across the screen in the 2014 Marvel Studios movie *Guardians of the Galaxy* as part of the cosmic collection of a character called The Collector.

Laika is the only nonhuman represented on the Monument to the Conquerors of Space at the Memorial Museum of Cosmonautics in Moscow. Positioned on the roof of the museum, the monument is a 350-foot-high titanium sculpture of a rocket leading a plume of exhaust and smoke from its engines. The monument is wrapped in a bas-relief of the heroes of the Soviet space program. I stood before the monument in the summer of 2016 and walked around its base. It was a particularly cool day, and it had only just stopped raining.

Through my reading I had come to know the names and faces of a number of the greatest figures in the early Soviet space program, but none of them was recognizable here. These were representations of a generalized Soviet hero, generalized faces in various generalized poses and actions. Laika, however, was recognizable, sitting in the capsule that took her to the stars. She is positioned over the shoulders of a man who is kneeling and looking at a set of drawings or plans, and beneath the raised arm of another man who, along with the rocket behind him, points the way to the stars. That monument—on which Laika is a central figure—writes Olesya Turkina in *Soviet Space Dogs*, "came to symbolise the hopes and dreams of an entire generation of Soviet people."

In the West, too, Laika lived large in the public imagination. For many, she was proof that the Soviet Union now possessed the power to destroy the United States, or any other country for that matter, from space. And she was a symbol of Soviet godlessness, a cruel sacrifice to the technological advance of a state fallen from God's grace. The scientific achievement of the Soviet Union became lost in the story of Laika herself, as a global debate broke out about the ethical use of animals in research, fueled primarily by anti-Soviet propaganda from the National Canine Defense League (now Dogs Trust) and the Royal Society for the Prevention of Cruelty to Animals, both in the United Kingdom. Other animal rights organizations joined in to encourage protests at Soviet embassies worldwide. No matter that, worldwide and in the West where these protests originated, animals were likewise being used in medical and scientific research. To keep pace with the Soviets, the US military was sending not dogs but monkeys, and eventually chimpanzees, into space in experimental rockets, many of which crashed and burned up on the desert plains of eastern New Mexico. But most people do not care about monkeys and chimps the way they care about dogs, and so Laika—a stray dog in a rocket ship, dead and exiled, orbiting the Earth for months—became a symbol of the USSR's power and technological superiority, and in the West a symbol of that power's cruelty and indifference to the value of life, human and animal.

Despite the outcry in the West against the Soviets for their treatment of Laika, she was beloved by the team working on *Sputnik II*, as were all the space dogs in the program. These were the days before standards and laws governing the treatment of animals in labs, but even without such laws, the Soviets took great care of their space dogs as well as the other animals they used in scientific research. Stories of the Soviet scientists' and engineers' attachment to and affinity for the dogs abound, Laika chief among them. In a 2006 telephone interview with Chris Dubbs, co-author of *Animals in Space*, NASA's chief veterinary officer, Joseph Bielitzki, spoke about the way Russians work with animals, in the past and the present. Russian animal care programs, Bielitzki said, were "more personal than [the US]," adding that "you can't work with the same animals for almost two years, as they did, without becoming emotionally attached to them." Vladimir Yazdovsky broke strict regulations to take Laika home to play with his children not long before her scheduled launch. In his memoirs he writes that he "wanted to do something nice for the dog since she didn't have much longer to live." And in 1998 Oleg Gazenko, who worked under Yazdovsky, expressed his sadness and regret for sending Laika to her death: "Work with animals is a source of suffering to us all.... The more time passes, the more I am sorry about [Laika's death]. We did not learn enough from the mission to justify the death of the dog."

But it was the death of Laika that positioned her so firmly in the cultural memory of a generation, not just in the Soviet Union but in the United States, the United Kingdom, and many other countries. While her story raises questions about the ethical treatment of animals in research, it may in fact have helped change the way we think about such animals, and the way we treat them. Laika may have helped change the way we think about the Space Race. Cathleen Lewis, a curator at the Smithsonian's National Air and Space Museum in Washington, DC, told me that in the 1940s and 1950s "we were just stumbling into the space age," and the space age was a highly mechanized and technological endeavor. Then "we sent a living being into space. It wasn't an anonymous cage rat, but a dog. Dogs are essential to human evolution. They taught us a lot about our social behavior. To have an animal like a dog sent into space really shifted the Space Race, gave it a less military, less threatening face."

While Laika may have made a highly mechanized and technological endeavor more human, she also became inseparable from it. Without her capsule to provide oxygen, proper atmospheric pressure, and temperature control, Laika would not have survived in orbit for any length of time. That, along with the sensors implanted in her body and attached to the satellite for data transmission back to Earth, we have in Laika the union of the biological and the technological. The satellite was built around the dog, not the other way around, and in it she became part of it, just as it became part of her. Its fate was bound to her fate, satellite and dog.

Within three years after Laika's flight, NASA scientists Manfred Clynes and Nathan S. Kline coined the term "cyborg" to describe a biological organism integrated with artificial components or with technology. The purpose of such an integration, as they saw it, was not to make the biological organism more like a machine but to make it a better version of itself. In their 1960 paper "Cyborgs and Space," Clynes and Kline asserted that instead of building complex environments inside which humans might travel in space and live on alien worlds, the better path was to alter humans so that we might manage such environments. Integration with technology would free us from it, they surmised, allowing us "to explore, to create, to think, and to feel." In altering the way our physical bodies work and the kinds of environments we can tolerate, we would alter too the meaning of being human. We would broaden and expand what we call the human spirit. "Space travel challenges mankind not only technologically but also spiritually," Clynes and Kline wrote, "in that it invites man to take an active part in his own biological evolution."

We have become so dependent on technology in the twenty-first century that we can hardly get through a day without it. We have developed artificial components for the body: knees, hips, pacemakers, heart valves, and in some cases computer chips. We are attached to our automobiles, which are increasingly driven by computers and integrated with our handheld computers, our mobile phones. We are inseparable from those mobile phones, which serve our social, economic, political, and educational needs. Our phones are our identity, or at least our identity—who we say we are, and who we wish to be—is expressed through our phones. We are so integrated with our technology that it may soon be appropriate to call ourselves cyborgs. Laika in her capsule may stand as an origin point and as a symbol for this new way of being in the world, this new way of being in the cosmos.

Laika's story also raises questions about human empathy, about what we are willing to risk in the spirit of science and exploration, both of animals and of ourselves. It offers a glimpse into the hearts of the Soviet scientists and engineers who worked with her, and into the hearts of us all. The scientists' relationship with Laika tells us something, perhaps, of what has changed for us in the years since *Sputnik*. What does it mean, for example, that Cold War Soviet scientists were not inhumane and heartless, as Western protesters insisted and the press reported, but were, in fact, caring, sensitive, and empathetic?

The philosopher Emmanuel Levinas tells us that it is in face-to-face relationships that empathy, and thus ethical behavior, becomes possible. To face another human being, to encounter them, is to come into an ethical relationship with them. Ethical behavior toward other human beings extends, it seems to me, to ethical behavior toward other species and toward the whole of nature. In this light, it is essential, I think, to consider the way we communicate and build personal relationships with other human beings in the twenty-first century. Even as our technologies enhance us, possibly make us better, we must also ask what we may have lost. Our obsessive use of communications technologies—chiefly smartphones and computers with internet access—has brought the world together, but it has also separated us, cutting us off from the kind of face-to-face encounters that Levinas writes about. What are the results of this recent phenomenon? What is its future? How do these technologies advance us as we become more and more dependent on them? How do they limit us?

With all these considerations, Laika really belongs to the world. She was a Russian dog with a Russian heart, a stray from the streets of Moscow, a creature of the Soviet experiment, but once lofted into space she was not fixed or bound to any one place, to any one people. She belonged to no one and to everyone, the first living creature from Earth to go where no nation, no corporation, and no individual has any claim. In looking at Laika's story and at the scientists and engineers who worked with her, perhaps we can better understand human empathy—its source, its importance, and the dangers of its reduction and loss—in order to posit that empathy is essential to caring for our world, to maintaining it as a place where we may all live healthy, productive lives as part of what the writer Wendell Berry has called "the feast of creation."

◻

When the Soviet team went to select a space dog to launch in *Sputnik II*, Laika earned high marks. In her training she managed the extreme conditions of the centrifuge and the vibration table, and she kept a calm and even disposition during prolonged periods of isolation in the training capsule (up to twenty days). She did not become aggressive or fight with her kennel mates, as so many smaller dogs are prone to do. The women and men who worked with her describe her as sweet-natured, patient, and determined, a dog that wanted to please, a dog talented in adapting to the place and the people with whom she found herself. She was a survivor, a quality that saw her through that hard life on the streets and, then, the rigorous training in the space dog program. Indeed, Korolev and Yazdovsky both knew that street dogs made the best space dogs, because they were tough, scrappy, and they could endure extremes of temperature, hunger, and isolation.

And yet Laika was not the most qualified dog for *Sputnik II*. The team felt strongly that a dog called Albina was the best choice. Albina was a favorite among the scientists and engineers, "a celebrity who had twice been in research rockets at the height of hundreds of kilometers," writes engineer Oleg Ivanovsky (under the pen name Aleksei Ivanov) in *The First Steps: An Engineer's Notes*. Albina had flown both times with a dog called Kozyavka, or Little Gnat, in June 1956. Having already proven herself in flight, she was the perfect choice for *Sputnik II*, but had she not risked enough? Didn't she deserve something for the contribution she had already made? Retirement, perhaps, a soft bed in a warm house? Albina had something else going for her too: she had just given birth to a litter of puppies, three little pups, one of which looked a lot like Laika. Yazdovsky thought it too cruel to take the mother from her pups and subject her again to the risks of rocket flight.

Ordinary rocket flight was risk enough, but *Sputnik II* was not going to be an ordinary flight. Khrushchev had ordered the launch for no later than November 7, 1957, in time to celebrate the Bolshevik Revolution, which was about a month after *Sputnik I*. The Soviets knew how to send a dog to the edge of space on a rocket and bring it back safely with a system of parachutes (they had achieved this many times), but they did not know how to return a dog from orbit. It had never been done before. The know-how and hardware just didn't exist. That kind of research and development took time, and the Soviet team was in a mad race against the clock, and against the United States, to again achieve the impossible. So unlike previous and later rocket trials using dogs, the dog chosen to fly on *Sputnik II* was not just taking a risk, it was never coming back. The dog chosen for *Sputnik II* was going to die in space or perhaps on the ride into space. Whatever, it was going to die.

So which dog to choose: Laika or Albina?

For the team, it was a tough choice, but a choice had to be made. Ivanovsky records this moment in *The First Steps:* "The great majority inclined to send Laika into space. Everyone knew that the animal would die, and there was no way to bring her back to Earth, because we did not know how to do that. So it was particularly painful to send Albina, everyone's darling, to her death. Thus, Laika became the first." By the first, Ivanovsky means "the primary," as all the space dogs scheduled to fly were assigned a second, or a backup, in case something happened to prevent the primary from flying. Albina, then, was named Laika's second. Ivanovsky's words also mean that

Laika would be first into orbit, first for the glory of the Soviet Union, and first for all time. And she would also be the first, and the only, dog in the Soviet space dog program to be sent to her death.

▢

Laika's capsule on *Sputnik II* contained a small, round window. While the satellite, along with Laika and her capsule, were destroyed on reentry, the window is plainly visible in photographs and drawings, and in a replica at the Memorial Museum of Cosmonautics in Moscow. The window, Asif Siddiqi told me, measured about 6.3 inches in diameter and was made of transparent thermoplastic, called Perspex. Questions surrounding the window's purpose stand as a lodestar for this book, out of which a central metaphor has arisen. Wearing a flight suit to secure the sensors implanted in her body, Laika was positioned inside a small chamber within the capsule so that, while she was able to stand, to sit, to lie down, and to move about a little, she was unable to turn around. The design meant that she faced the window looking out. Was this the sole reason for the window, so that Laika could look out at the Earth as she left it? Or did scientists believe that the view or, more so, the entry of natural light would help calm her as she endured the thrust of the rocket and then the strange sensation of weightlessness? Or was the window installed to serve the scientists only, to monitor Laika before she was loaded onto the rocket? What were these scientists' intentions? After launch, of course, the sole user of the window was Laika, a window through which she looked at what? Did Laika take in the first view of Earth from space and look beyond it into the cosmos? Could she in fact see out? And if she could, what did she see from up there? What did she see down here? Or was the view from Laika's window occluded by darkness, blocked by the payload fairing protecting the satellite as it rose into orbit? More specifically, did the Soviet team remove the payload fairing after *Sputnik II* achieved orbit?

Remember the famed 1972 photograph of the first full view of the Earth taken from the American spacecraft *Apollo 17* on its way to the moon? That single image, known as the Blue Marble, is one of the most reproduced photographs in human history. Why? Because it, more than any other image, brings human beings face-to-face—and for the first time—with the vulnerability and special quality of our planet home. Because that photograph allows us to see, to witness, even to feel that the Earth is fragile, and its fragility makes it beautiful. Because that image brings us all to square with what we have always known: that we have but one home, one Earth, and we must take care of it, if we are to survive. "[The Earth] truly is an oasis," said *Apollo 15* commander Dave Scott in an interview for the documentary *In the Shadow of the Moon*, "and we don't take very good care of it. I think the elevation of that awareness is a real contribution to saving the Earth." It's a jewel, he said, the "jewel of the Earth hanging in the blackness of space."

On April 12, 1961, some four years after Laika's mission, the Soviet Union put the first human being into orbit, cosmonaut Yuri Gagarin. During his single orbit of the Earth, Gagarin gazed back on our planet home. He could not see the Earth entire as the Apollo astronauts would as they sailed away from it, but he was the first human being to look down from up so high. "Orbiting Earth in the spaceship, I saw how beautiful our planet is," Gagarin later wrote. "People, let us preserve and increase this beauty, not destroy it." Gagarin's is a necessary plea, and further evidence that seeing the Earth from space changes the fundamental way we think of ourselves, making it difficult to see ourselves as superior to other living beings, as the inheritors of a planet filled with resources and placed here for us, as made in the image of a benevolent god. Seeing the Earth from space, we must face a terrifying reality: that we are fragile, vulnerable, and possibly alone. What shall we do with this new point of view? How shall we account for who and what we are, and for the fragility and essentiality of our planet?

Did Laika see this sight too? Did Laika see it first? And what value is there in knowing, or not knowing, that she did?

Whether she could or not, by imagining the view from Laika's window—from the inside of her capsule looking out—perhaps we can learn more about her and more about ourselves too: who we are as a species, with our appetite for exploration and knowledge, our appetite for power too, running side by side with our compassion, our empathy, and even our love for the planet we live on.

And Laika's story is about loneliness, about loneliness as fundamental to being human. If we can trust poet Sylvia Plath, loneliness is an assertion of our need for each other, our need for "another soul to cling to." In clinging to another soul, we do find comfort, but we lose that comfort to partings, breakups, wars, death, as if the universe itself is so hinged to cycle us through loneliness and comfort and then loneliness again like seasons turning across the years. These seasons of losses are the linchpins of all this loneliness, loneliness no human being can escape. In response to it, we cast our net into the cold, black cosmos, our net of satellites, telescopes, rovers, probes, spacecraft, space stations: is anything out there? Is anyone out there? And we send our dogs out too, out into the void to help us with the question we all hope to answer: are we alone in the universe? Are we special, or are we instead rather ordinary because life abounds out there? Will it one day be possible to establish fellowship with other beings on other worlds? Whether we will or not, human life is without meaning without one another, and without animals; all we have is each other. As the novelist Kurt Vonnegut stated in a 1974 commencement address, "The most daring thing is to create stable communities in which the terrible disease of loneliness can be cured."

Laika's story, then, is important because it draws us into an ethical relationship with animals we depend on, specifically with dogs as our companions in cosmic exploration. We are operating here in our search of the heavens at the very limit of our technological capability, and right beside us, with Laika as its apologue, is the dog, the animal with which we share a close evolutionary history. Wherever human beings go, whatever human beings do, our dogs go and do it with us. Sojourning with our dogs on the hunt for answers, even across an ocean of stars, is itself a kind of comfort. If adventure and exploration are evidence of an innate loneliness in the human animal, they are also an antidote to that loneliness. Space exploration, ultimately, may be a search for a cure for loneliness.

Finally, at its core, this book is a story about a dog. It is a biography of sorts, a portrait, a memorial. This book belongs to Laika and to all the space dogs who traveled before and after her. Through the stories of their journeys into space, perhaps we will all find ourselves a little more grounded here on Earth.

○

In order to understand Laika's story, it is important to come to know some of the other animals that have flown into space, who sent them and why, and what benefit, if any, resulted from their missions. It is also essential to acquaint ourselves with the stories of the other space dogs, those that flew before Laika, and those that flew after. Chapter 2, "Animals in the Heavens," traces the story of animals in space from the earliest documented flights in hot air balloons in the eighteenth century to the animals that are part of the ongoing science aboard various satellites and on the ISS. Since the 1970s, robotic rovers and probes have mostly replaced nonhuman animal explorers and have been loosely remade in their image. While this book is not an indictment of research using animals, and it is especially not an indictment of the Soviet scientists, or any others, who launched animals into space, it does ask questions about the value of such research weighed against the suffering of the animals.

Chapter 3, "The Making of a Space Dog," returns to the story of Laika to detail her training and includes generally the training that all Soviet space dogs endured. It also introduces

Sergei Korolev, the Soviet chief designer, whose single-purpose life serves as a balefire for humanity's entrance into the Space Age. This chapter prepares the way to better understand the specific details of Laika's flight on *Sputnik II.* Chapter 4, "Scouting the Atmosphere," features the stories of the most notable space dog flights before and after Laika. While a few of these flights rival Laika's for achievement in endurance and advancements in technology, it is clear that she serves as a tipping point in space exploration, beyond which the dream of exploring nearby and distant planets opened into a kind of fever from which humanity has never recovered. Now that we know space travel is possible, it seems to me, we are no longer content with living solely on Earth. This chapter also brings us to the conclusion of Korolev's story and positions him as the man who matched, if not exceeded, the achievements of the greatest minds in space exploration. Like Laika, he will forever be listed among those who did it first and those who did it best.

With these preparations, chapters 5 and 6, "A Face in the Window" and "First Around the Earth," tell the story of Laika's flight. In these pages I have assembled—and I think for the first time—an intimate portrait of Laika: where she came from, what she endured, and what her flight means for us all. I use what is known to imagine what cannot be known: Laika's experience inside her capsule during her time in orbit, for example. Here my purpose is not melodrama or embellishment but witness and understanding. In addition, my research brought me to an as yet unacknowledged truth about Laika's flight and death, which I hope aids our understanding of her and our understanding of ourselves, which is this book's main premise.

Finally, the epilogue addresses what comes next in human space exploration and draws a straight line from the aspirations of the earliest rocket scientists in the late nineteenth century, who dreamed of traveling to Mars, to Robert Zubrin and the Mars Society and Elon Musk and his SpaceX in the twenty-first century. While Laika is the Earth's first space traveler, she was also one scout among many, sent out ahead to gather information to help us all on our journey to the stars.

TWO

○

Animals in the Heavens

All the universe is full of the life of perfect creatures.

KONSTANTIN TSIOLKOVSKY
"The Scientific Ethics," 1930

While Laika was the first living being in orbit, she was not the first living being in space. By the time of Laika's flight, both the Soviet Union and the United States had been experimenting for a decade with suborbital rocket flights carrying dogs and monkeys, respectively, but also fruit flies, mice, and other living things, just beyond the Karman line (100 kilometers or about 62 miles altitude) that marks the boundary of space. It was the Hungarian American physicist Theodore von Karman who first calculated that at this altitude, the atmosphere becomes too thin to support traditional flight. To navigate, and even to survive, a pilot needed a wholly new kind of craft, not an aircraft at all but a spacecraft lifted into the heavens on the nose of a rocket.

Some of these early experimental rockets carrying animals blew up on the launchpad, some turned and tumbled in the sky and blew up, some flew erratically and strayed off-course and were destroyed by charges inside the rocket detonated by a ground crew, and some rockets flew up and up into space, where they sojourned in microgravity, then turned a delicate arc back to Earth, the braking chutes deploying and slowing the spacecraft for a landing on the ground. And when the scientists and engineers arrived to retrieve those spacecraft and their animal passengers, sometimes they found them dead and sometimes they found them alive. And all of it in pursuit of science to reveal the mysterious conditions of space, and to one day give them confidence enough to risk sending the first human being off the planet. Human spaceflight, human voyages to nearby and distant planets, has always been the goal, even from the late nineteenth and early twentieth centuries when the first spacecraft designs took form in the minds of a few visionaries from Russia (later the Soviet Union), Germany, and the United States, among other countries. *Sputnik II*, carrying Laika, was neither the beginning nor the end. It was one step along a winding path in the exploration of the final frontier, but a giant step to be sure, one for the record books.

Since those early rocket flights in the Soviet Union and the United States, five more nations have flown animals in space: France, Argentina, China, Japan, and Iran. The list of the kinds of animals, and also plants, sent into space is dizzying and includes quail and quail eggs, butterflies, mollusks, various fish, including the mummichog minnow and the oyster toadfish (a hideous beast at best), various mosses, oat and mung bean seedlings, newts, worms of all sorts, and nematodes, which come in all sorts too. Cats, rabbits, rats, Mad-

2016 NASA steered the Mars rover *Curiosity* away from possible water sources on the red planet so as not to contaminate them with Earth microbes.

Other objects, too, surprising and strange, have been sent into space, among them the light saber Luke Skywalker used in *Return of the Jedi*, images of Playboy models secreted away in the task notebooks of *Apollo 12* astronauts, a wheel of Gruyère cheese, dinosaur bones, Coke and Pepsi, a Pizza Hut pizza (literally the first out-of-this world delivery), a Buzz Lightyear toy, Amelia Earhart's watch, and samples of the remains of the late Gene Roddenberry (creator of *Star Trek*), James Doohan (who played Scotty in the original *Star Trek*), and astronomer Clyde Tombaugh, whose sample of remains reached Pluto on July 14, 2015, on NASA's *New Horizons* probe, because the man discovered it.

❑

As far as anyone knows, the first animal test flight in the history of the world took place on November 19, 1783, at Versailles. This is, of course, not including the certainty that for the past two hundred thousand years, boys have tested the reliability of gravity by cruelly launching little animals off high places. The Montgolfier brothers, Joseph and Étienne, were not cruel boys, however, but young inventors and entrepreneurs with talent in mechanics and science who had been experimenting with hot air balloons for about a year. They arrived at this work, so the story goes, when Joseph, the starry-eyed dreamer with a reputation for reciting Voltaire, was drying his wife's chemise by the heat of a wood fire. Filling temporarily with hot air, the garment billowed up. In his excitement, Joseph's mind began to wander: could not a sack or balloon of some kind fill with such air and so be borne aloft? If so, could not a balloon of greater size be constructed, filled with the heat of a fire, and so bear something of weight, a man, for example, up into the sky? With such a balloon, Joseph wondered while clutching his wife's chemise, was it possible that a man could fly?

Another account has Joseph gazing at a painting of the siege of Gibraltar, that three-year failed attempt by Spanish and French forces to take Gibraltar from the British during the American Revolution. The siege was even then under way, and Joseph, a good Frenchman, his mind bent on aiding his country, recalled a past moment watching sparks drawn up a flue. He wondered if men, like sparks, might be drawn up on a draft of air from a fire, up and over the castle walls, thus impregnating impregnable Gibraltar. Flight has always been associated with military conquest.

It was time to perform an experiment. In his rented rooms in Avignon, Joseph stretched a bolt of taffeta around a light wood frame and lit a wad of paper in the cavern of it, and the whole thing floated up to the ceiling. As Charles Coulston Gillispie reports in *The Montgolfier Brothers and the Invention of Aviation, 1783–1784,* Joseph immediately wrote to his brother, Étienne: "Get in a supply of taffeta and of cordage, quickly, and you will see one of the most astonishing sights in the world."

Luckily the family business was papermaking, a high-tech industry in the eighteenth century. The brothers set to work experimenting with paper balloons, which became known as Montgolfières, and the hot air that inflated them as Montgolfier gas. Gas, because the brothers had yet to understand that the fire heated the air, and that hot air is lighter than cold air, so it rises. Instead they concluded that the combination of fire and smoke caused a chemical reaction with the air and the resulting gas filled the balloon. No matter how it worked, it worked, and so the brothers partnered with several interested friends, including Jean-Baptiste Réveillon, a successful manufacturer of wallpapers. Surely, writes Gillispie in his book, it is due to the association of Réveillon with the work of the Montgolfiers "that the iconography of balloons has evolved out of the patterning of eighteenth-century wallpaper."

On that day in Versailles, the brothers planned a public demonstration. They had already put claim to their invention (and so the invention of aviation) with a previous public demonstration, which launched from Annonay on June 4, 1783. But this flight in Versailles would be different. First, the Montgolfier brothers had to contend with the success of another ballooning pair, Jacques Alexandre César Charles and Barthélemy Faujas de Saint-Fond, who had only a month before launched a helium-filled balloon from the Champ de Mars. Second, the brothers were to stage this launch before the Royal Palace and before royalty—namely the king, Louis XVI, and his queen, Marie Antoinette (not so much later, both would lose their heads to the guillotine). And third, the brothers would raise the stakes by flying a living animal. After some deliberation on what animal to fly (some suggested a dog, so that while ascending into the sky, the crowd of onlookers could hear it bark), the brothers settled on a sheep, a duck, and a rooster. The duck and the rooster were control animals for the sheep. The duck was expected to manage the altitude without trouble, but what would happen to the rooster, no one knew. And what happened to that sheep during the flight, a mammal with a physiology not unlike human beings, would be akin to what would happen to a man, so the thinking went. If the sheep came down no worse for wear, then the brothers would most certainly make plans for a manned flight. They called the sheep Monteauciel, meaning "ascend to the sky."

A great crowd gathered at the Royal Palace. All the chateau windows and the rooftops were crowded with onlookers. The balloon had been constructed hastily, in just four days' time, when the intended paper balloon—a beautiful thing measuring seventy feet high by forty feet wide, with a background colored in azure and ornamented with gold in representation of the sun—was destroyed by rain. For this new balloon, the brothers had turned again to taffeta, that crisp, smooth silk fabric, and coated it with varnish against all weathers. It was not as big but equally stunning, a bright blue bulb ribboned by two golden bands.

These first flying balloons were not constructed with an onboard heat source; they were filled and then released. The brothers filled their balloon with Montgolfier gas, and at the signal, the dozen or so men holding it back on tethers all let go at once. "The machine rose majestically," Étienne wrote to his wife, Adélaïde, "drawing after it a cage containing a sheep, a rooster, and a duck." Then a gust of wind tilted the balloon at a sharp angle, spilling some of the hot air from the bottom now too steeply pointed up, but it soon righted itself again. It "continued on its way as majestically as ever for a distance of

1,800 fathoms [about two miles] where the wind tipped it over again so that it settled gently down on the earth" at the edge of the forest of Vaucresson. Upon inspection, the balloon had sustained some damage in the upper reaches of its curve, but the animals, Étienne reported, "were in fine shape, and the sheep had pissed in the cage."

Shortly after this historic flight, the Montgolfier brothers constructed a balloon twice as big to raise a man into the sky. Certainly there was no hope that the balloon would reach space (which is, in fact, not possible for any balloon because at high altitude, the air becomes too thin to keep it inflated), but just to get the thing up and then safely down was an achievement beyond compare. The honor of becoming the world's first pilot is most always granted to the twenty-six-year-old showboater François Pilatre de Rozier, who offered himself up as a test subject and made several flights in a Montgolfier balloon. But it is nearly certain that it was Étienne himself who was first, riding a balloon into the sky from the yard of Réveillon's wallpaper factory attached to the ground by a tether.

○

A few years after the Mongolfiers, Claude Ruggieri, an Italian living in Paris, started messing around with rockets powered by gunpowder. He was one of many such curious people romanced by the flash and bang of explosions, and his name rose to prominence as early as 1806 when he began staging public demonstrations. He loaded his rockets with mice and rats and returned them to Earth under a parachute. His family had long been in the fireworks business, and launching small animals into the sky was a logical next step. By 1830 Ruggieri was building larger and larger rockets, and he announced that he would launch a ram into the sky from the Champ de Mars. Surely he knew about the Montgolfier brothers, and with this flight perhaps he would match them. The Eiffel Tower had not yet been constructed, but the location was no less dramatic. Pitched by the excitement of the spectacle, a young man volunteered himself in place of the ram. Ruggieri accepted. Moments before launch, the French police arrived to cancel the demonstration. As it turned out, the man was not a man at all but just a boy eleven years old.

◻

While the Montgolfier balloon flight was the first with animals, it certainly was not the last. Balloon flights have persisted as a relatively inexpensive and efficient way to send animals, and so science, high into the atmosphere. In 1862 Britain's famed meteorologist Sir James Glaisher and Henry Coxwell ascended to about 35,000 feet in a balloon they called *Mars*. They carried with them a cage full of pigeons and marked various altitudes by casting them over the side, one by one. Pigeons can fly up to about 6,000 feet altitude, but as the balloon got higher and higher, the birds became dopier and dopier until they could not fly at all. Now when the men released them, they fell leadenly away and out of sight. Glaisher passed out at about 28,800 feet, and Coxwell, in a hypoxic stupor, saved both of their lives by releasing air from the balloon, allowing them to descend to a lower altitude. It is not known what became of the pigeons.

Almost one hundred years later (1947–1960), the US Air

Force experimented with high-altitude balloon flights carrying animal passengers out of Holloman Air Force Base, adjacent to the army's White Sands Missile Range in New Mexico. Holloman launched balloons carrying fruit flies, mice, rats, hamsters, guinea pigs, chickens, rabbits, cats, dogs, frogs, goldfish, monkeys, as well as fungi and various seeds. What happened to these animals at high altitude would happen to a human being, so the logic went. The resulting data was then used to develop life-support technologies that would allow humans to eventually endure the conditions up high.

In the US, Holloman and White Sands are ground zero for balloon test flights as well as rocket and missile test flights, rocket test flights carrying animals, and nuclear tests. While nearby Roswell, New Mexico, has built a tourist industry around the supposed 1947 crash landing of an alien spacecraft, the US military has confirmed that what really came down on that June day was a high-altitude balloon sent up to detect sound waves from Soviet nuclear tests. Even as the Soviets would dominate the Space Race at least until the first *Apollo* moon landing in 1969, the United States could claim detonation of the world's first nuclear bomb on July 16, 1945, at the Trinity Site on the north end of White Sands. That test led to the bombs that destroyed Hiroshima and Nagasaki and forced Japan's surrender at the end of World War II.

NASA still uses high-altitude balloons in scientific research, although they are now mostly uncrewed and un-animaled, as evidenced by the 2016 flight of their Super Pressure Balloon launched from Wanaka, New Zealand. The balloon remained aloft just shy of forty-seven days, setting a flight duration record at midlatitudes. Over its thirty-five years of operation, NASA's scientific balloon program has launched more than 1,700 balloons.

◻

The fruit fly—that tiny fly with red eyes and a black-and-tan striped abdomen—is likely the first animal to ascend beyond the Karman line into space, which it did on a V-2 rocket on February 20, 1947, out of White Sands. German scientists developed the V-2 as a missile for Hitler's final push toward the end of World War II. After the war, Russia and the US swept in and seized the V-2 hardware and the scientists for their own. The experiment on this particular V-2 was set up to test the effects of radiation at high altitude, one of the primary areas of research in the safety and even the feasibility of human spaceflight. You send up a bunch of fruit flies and see what condition they are in when they come back. That V-2 rocket rose to an altitude of sixty-eight miles, and then its little capsule, called the *Blossom*, returned to Earth on a braking chute. The fruit flies were recovered alive, and ever after, fruit flies have been our continual partners in biological research in space. Fruit flies have ascended into the upper atmosphere on balloons, on space shuttle flights, on most of the various space stations, and they have gone out and come back on biological satellite flights.

Humans owe a great debt to the ordinary fruit fly, because it does the heavy lifting when it comes to the study of genetics, both on Earth and in space. "The fruit fly has turned out to be a workhorse organism," said Jeffrey Thomas of the Department of Cell Biology and Biochemistry at Texas Tech University's Health Sciences Center. "From the 1940s through the 1960s no other animal was their equal in biological research. They're underappreciated, I would say. We really wouldn't be where we are today in biology and medicine without them." Our understanding of basic genetics, the effects of radiation on heredity, the role of chromosomes in heredity, embryogenesis, the role of cell communication in disease—all this we owe to the fruit fly.

While fruit flies have long been part of short-term experiments in space, in 2014 NASA established the Fruit Fly Lab on the ISS to begin long-term experiments. In particular, the lab will study the effect of microgravity on fruit flies, which will help scientists understand such effects on humans. Additionally, NASA says, fruit flies will help us understand "the effects of spaceflight on the immune system, the development cycle (birth, growth, reproduction, aging), and behavior."

⬚

A decade before Laika, the US Air Force at White Sands flew a series of monkeys into space for the Albert Project, which began with the flight of Albert I in 1948 and concluded with Albert VI in 1951. Weighing eight to ten pounds each, all of the Alberts were rhesus monkeys, except for Albert III, who was a cynomolgus monkey. Secured in protective harnesses and loaded into sealed capsules, the monkeys were anesthetized to avoid discomfort during flight. Albert I's capsule was so cramped that when the monkey was inserted its head had to be pushed down, bending its neck at an acute angle. Before launch, someone wrote on one of the rocket's fins, "Alas, poor Yorick. I knew him well," a misquote from Shakespeare's *Hamlet*. In the control room, the team reported that no heart rate or respiration registered on their instruments. The monkey likely died of asphyxiation, they concluded, from its bent neck. The rocket went up anyway, carrying a dead monkey, and reached an apex of thirty-seven miles. On the descent, the braking chute shredded at 25,000 feet, and the spacecraft broke apart on impact, leaving a waste of wreckage on the desert floor. Later the team confessed that they were unable to retrieve any data, but they learned something about the V-2 rockets they were working with and about capsule recovery. They would try again.

With each successive Albert flight, the resulting data led to improvements in capsule and braking design and instrumentation, but they were all failures as far as the monkeys were concerned. Albert II died when the braking chute on his spacecraft failed and he slammed into the Earth. The impact carved out a crater ten feet across and five feet deep. Albert III died when his rocket exploded during flight. Albert IV's braking chute failed, and the capsule crashed. While Albert V flew in a new rocket design, the more reliable and higher-performance Aerobee rocket, again the braking chute failed. The rocket crashed and was lost in the desert. Eleven mice joined the flight of Albert VI, some for testing microgravity and others for testing the effects of radiation exposure in space. Albert VI went up and came down, and the chute system worked. He landed safely back on Earth. The recovery team didn't arrive for two hours, and, trapped inside the capsule, Albert VI died of heat exhaustion in the New Mexico sun.

On December 3, 1958, newly formed NASA loaded a one-pound squirrel monkey named Gordo onto a Jupiter rocket. Gordo was also known as Old Reliable by the team that trained him because inside his little space capsule, he always fell asleep. Gordo wore a leather-lined plastic helmet and was strapped into a seat molded to his body. The rocket launched, pushing upward into the sky, Gordo enduring as much as a g-force of ten. At an altitude of about 300 miles, Gordo floated in microgravity for more than eight minutes, and on the descent the spacecraft hit 10,000mph. Instruments showed that Gordo survived all this, and in good condition, and if a man had been inside that capsule he would have survived too. The spacecraft splashed down in the Atlantic 1,500 miles downrange, but the recovery team was unable to locate it. After a six-hour search, it became clear that little Gordo would remain out there, forever lost at sea.

By 1959 the Soviets had put dozens of dogs into space (but not into orbit) and recovered them safely. The US had a couple of satellites in orbit, but with the exception of the fruit flies, it had failed every attempt at recovering a biological spaceflight. If the US was going to get a man safely into space before the Soviet Union, it had to get an animal safely into space first. And the US finally did with the flight of Able and Baker. Able was a rhesus monkey born in a pet shop in Independence, Missouri, and Baker was an eleven-ounce squirrel monkey from the jungles of Peru. The pair flew with a host of other biological experiments testing the effects of microgravity and radiation on living systems. According to Colin Burgess and Chris Dubbs in their seminal work, *Animals in Space*, these other passengers included "corn and mustard seeds, fruit fly larvae, human blood, mould spore and fish eggs, as well as sea-urchin shells and sperm, carefully triggered to produce fertilization during flight."

Able and Baker flew on May 28, 1959, reaching an apex of 360 miles altitude, where they drifted in microgravity for nine minutes. On the descent, the fantastic speed of the spacecraft subjected these two little monkeys to a crushing 38g. Able's heart rate went from 140 beats per minute to a peak of 222, and her respiration rate tripled. Baker, the tiny squirrel monkey, experienced a bit of "cardiac inhibition," reports call it, but her respirations remained normal. The spacecraft came down in the Atlantic, and navy frogmen helped hoist it onto the deck of the USS *Kiowa*. Able and Baker were unharmed, but not so for one of the recovery crewmen. As he was removing Baker from the capsule, she bit him.

Four days after her flight, Able died suddenly from a reaction to a mild anesthetic given to prepare her for the removal of the implanted ECG electrode. The anesthetic had been used routinely on hundreds of primates without incident. Her body was preserved, and she became part of the collection at the Smithsonian National Air and Space Museum in Washington, DC. Baker fared better, becoming a national celebrity and living a long life in retirement. She enjoyed an annual birthday party to celebrate her achievement and the newest advances in space technology, sometimes complete with a cake topped with bananas and strawberries. She made television appearances, and children wrote her thousands of letters. At the time of her death on November 29, 1984, she was considered the longest living squirrel monkey in captivity. She is buried at the US Space and Rocket Center in Huntsville, Alabama, where visitors sometimes leave bananas in memoriam.

◌

After the Soviet Union and the United States, France was the third nation to put animals into space. As early as 1949, France was at work constructing a launch facility at Hammaguir, a remote site in the Algerian Sahara, not far from its border with Morocco. Algeria was still a French colony in those days, and the location offered a high degree of secrecy as well as a test range free of human populations. From here, the French team sent three rats into space on separate flights—Hector (1960), Castor (1962), and Pollux (1962)—each fitted with an electrode harness surgically implanted in its brain. The plan was to measure brain activity (as well as other vital signs) in microgravity, with the aim of eventual human spaceflight. The rats wore linen flight suits, which allowed them to be suspended on wires inside tubes during flight to protect them from turbulence. Hector returned to Earth alive but was euthanized some six months later for dissection and study. Castor's flight went off course and landed thirty-seven miles outside the scheduled recovery zone. Seventy-five minutes after liftoff, the recovery helicopter finally located the spacecraft, but Castor was dead, a victim of the impossible Sahara sun. Pollux too was killed when his rocket strayed off course and came down somewhere in the empty desert, never to be found.

But when it comes to rockets, the French are best known for Félicette, the only house cat in the history of the world to fly in space. Félicette was a mostly black cat with white markings: white socks, white shoulders, and a white nose patch running up between her eyes. She had been picked up as a stray off Paris streets, where she had probably lived free on rats and handouts.

In the lab, Félicette was trained to endure the stresses of rocket flight. She was confined in a box and subjected to the simulated noise of a rocket's engines. She was turned in a circle, around and around, into a dizzying state. She was spun on a centrifuge to test her body's response to high g-force. And like the rats, Félicette was fitted with an electrode port surgically implanted in her brain. The port sat on the top of her head between her ears, cemented to her skull. The surgery took ten hours.

On October 18, 1963, the team dressed Félicette in a linen spacesuit and placed her in a metal restraining box with only her head sticking out. They attached an electrode cable to the port on her head, as you might plug an external hard drive into a computer. Then they inserted the cat in the box into a metal tube and sealed it up. File footage shows Félicette meowing as she goes into the tube. Meow. Meow. Meow. A steady rhythm with a steady rest interval. That small French rocket, the Véronique, rose to an altitude of ninety-seven miles, Félicette enduring an acceleration force of 9.5g. After separation from the rocket, her spacecraft came down in the sands of the Sahara. A helicopter brought out a recovery team. They opened up the spacecraft and pulled the restraining box out of the tube, and there was Félicette, still meowing.

◌

Microgravity and the ceaseless bombardment of solar and cosmic radiation were always going to be major challenges in space travel, but if a man could not survive the rocket flight and the spacecraft's return to Earth, there would be no space travel. And what if an astronaut had to eject from the spacecraft during his return? At what velocity and altitude was that ejection surely fatal, and at what velocity and altitude was it survivable? Only good research could help scientists and engineers answer these questions and develop technologies to manage the conditions of spaceflight. Such research would also be useful in developing safety equipment for aircraft and even for automobiles. Again, animals would be the first to test the limits of the body and of the equipment, and many of them would die doing it.

In the late 1940s through the 1950s air force colonel John Stapp and his team developed several kinds of rocket sleds that carried hogs, chimpanzees, black bears, and men along a track at very high speeds and then brought them to a sudden stop. Two of the most important measurements in these trials were g-force and jolts. Expressed using a lowercase "g" from the term "gravitational," g is a measurement of acceleration that increases the weight of an object while its mass remains the same. If a man weighs 170 pounds, enduring an acceleration of 10g increases his weight ten times. He would weigh, at least momentarily, 1,700 pounds. Jolts are a measurement of the rate of the change in acceleration or deceleration. Whether in a rocket, an aircraft, or a car, a sudden start or stop can be the difference between life and death, and where was that line?

How much could a man take? To answer these questions, Stapp ran trials on his rocket sleds, first at Edwards Air Force Base in California and then at Holloman in New Mexico, where he took command of the Aeromedical Field Laboratory.

Gil Moore, who knew Stapp, told me that his steel-rimmed glasses gave him the mild-mannered look of a high school teacher, but he was fearless. Stapp's policy was never to let one of his men ride a rocket sled unless he had ridden it first. In some twenty-nine rocket sled trials, Stapp reached a g maximum of 45, three times the suspected limit for the human body. He suffered a number of injuries, including hemorrhaged retinas, cracked ribs, and two broken wrists. Burgess and Dubbs write that, in addition to injuries like Stapp's, other volunteers sustained "abrasions, lost teeth fillings, concussion, and unconsciousness."

At Edwards, Stapp built the so-called Gee-Whizz machine, running trials with 185-pound mannequins or crash dummies. File footage shows all kinds of mishaps: the sled coming off the tracks, the restraining harness snapping and vaulting the mannequin hundreds of feet beyond the sled, the mannequin's head popping off. Chimpanzees rode the Gee-Whizz machine too, in various positions, and under anesthetic so that whatever happened to them, they would not know it. To simulate a plane crash, or perhaps a crash landing in a spacecraft, chimpanzees were positioned headfirst and lying down. Reports indicate that the maximum deceleration experienced by these chimpanzees was a top speed of 169mph stopping within eighteen feet, which resulted in a momentary g-force of 270, far beyond the g-force possible in an accelerating rocket.

Death was nearly certain. According to Burgess and Dubbs, one of the researchers working on the project described what became of the chimps after these tests as "a mess."

Anesthetized hogs were used too, strapped into rocket sleds in an upright position with a harnessing system. Working in the research repository and library at the New Mexico Museum of Space History in Alamogordo, I came across an air force file photograph of one of these hogs. A documentation card placed in the photo reads: "Project Barbecue, Run #22, 5 August 1952."

On the museum grounds is one of the decommissioned rocket sleds, this one called the Daisy Track. It was decommissioned in 1985 and later restored for display. The Daisy Track was powered by compressed air, not unlike the Daisy air rifle it is named after, and the sled was stopped by something called a water brake. If you walk along its length—now painted a bright aqua blue—you sense a vestigial drama that once unfolded here. At least four black bears were anesthetized and strapped into the Daisy Track to endure about 20g, and then euthanized and dissected as researchers hunted for internal injuries. In 1958 Captain Eli Beeding made a run on the Daisy Track with two albino rats. He faced backward, away from the direction he was going, while one rat faced forward and the other was strapped to something called an anti-g platform. The sled malfunctioned, coming to a stop more suddenly than planned and dramatically increasing the g-load. The rat on the anti-g platform came through just fine, which I suppose means the device worked. The rat pointed downstream on the track suffered but recovered. Beeding took a hit of 83g. Researchers later determined that had he been facing forward, the ride would have killed him.

On December 10, 1954, Stapp made his final test run, this time on a sled known as the Sonic Wind I. He was bound tightly to the sled so that no part of his body could move. Captain Joe Kittinger (who held the world record for the highest skydive until Felix Baumgartner broke it in 2012) flew a T-33 chase aircraft down the track line, and the sled pulled away from it. The sled hit a top speed of 632mph (nearly Mach 1) in five seconds and then came to a full stop in 1.4 seconds. Burgess and Dubbs write that it was the equivalent of "hitting a brick wall in a car travelling at 120 miles an hour." While the crew unstrapped him, Stapp noticed his vision was blurred. He thought perhaps he'd torn both his retinas and might be blind for the rest of his life. He recovered though, proving that a pilot could eject from an aircraft or capsule and likely survive.

◻

Not far from the Daisy Track on the grounds of the New Mexico Museum of Space History, I visited the burial site of Ham, the first primate in space. The museum is located on a hill at the western edge of the town of Alamogordo, the dry flank of the Lincoln National Forest running up behind it. Ham is buried before a sun-struck concrete marker adorned with blackened and desiccated bananas laid in by well-wishers and announcing the International Space Hall of Fame. But Ham (his name is a derivative of Holloman Aeromedical Research Laboratory at the nearby Holloman Air Force Base, where he lived and trained) has not been inducted into the hall of fame. Only human beings have received that honor so far. Nor even is all of Ham buried here, only the cremated remains of his skin and viscera. His bones lie in a drawer at the Smithsonian's National Museum of Natural History in Washington, DC, having been cleaned by a colony of flesh-eating dermestid beetles.

After Laika and the space dogs, Ham is probably the most famous animal to fly in space, along with his counterpart Enos. It was Laika's flight that brought newly formed NASA to Holloman asking to put a chimp into space. If the Soviets could fly a dog into orbit, surely the Americans could fly a chimpanzee. And as our closest relative, the chimpanzee is the perfect test animal to prepare the way for the first human spaceflight. The US still had a chance to put up the first man, and to put up the first man and not kill him you had to test the equipment on an animal you could kill. Holloman already had a chimp colony for research, and the air force began to train some forty chimps for spaceflight. Ham and Enos emerged as the best of the best. They would test the rocket and capsule life-support systems for Project Mercury's seven astronauts in training, whose story was made famous by Tom Wolfe's book and the subsequent movie *The Right Stuff.*

Ham was born in equatorial Africa sometime in July 1957. Captured by animal trappers, he came into possession of the Miami Rare Bird Farm, a kind of breeding facility and tourist attraction (now defunct) from which the air force purchased him for $457. He was a wee chimp when he came to Holloman, weighing in at 19 pounds. Compare that to his flight weight of 37 pounds four years later, then the heaviest animal shot into space, and his gargantuan 175 pounds in retirement, first at the National Zoological Park in Washington, DC, and then at the North Carolina Zoological Park, where in 1983 he died of liver failure and an enlarged heart.

By the late 1950s plenty of animals had been shot into space and brought back alive, proving that living organisms could manage increased radiation and microgravity, at least in the short-term. What NASA did not know was if a man could work in space, and if he could perform the tasks required to operate a spacecraft. In addition to testing the life-support systems of the *Mercury* capsule, sending a chimp into space would help answer this question.

Ham and Enos were both trained to operate a series of dummy levers inside the *Mercury* capsule. Strapped into a chair, the chimps sat before a control panel with three lights, each with a corresponding lever. Burgess and Dubbs report that on the far right was a red light that glowed all the time, indicating that the chimp should not press this lever. In the middle was a white light that switched on when the chimp pressed the lever, which he was trained to do every twenty seconds. On the far left, a blue light came on unpredictably during an interval of two minutes, prompting the chimp to press the lever. If the chimp made an error, he was given an electric shock through a metal plate to which his feet were strapped. Because the chimp did not want to get shocked, he did not want to make a mistake. If a chimp could ascend into space on a rocket and return safely to Earth while performing these tasks with the levers, the research team reasoned, then a man could operate a spacecraft on a journey into space.

The chimps were trained by a team led by Master Sergeant Edward Dittmer, an aeromedical technician who began his career in the US Army in 1943 and then transferred to the air force in 1947. In 1955 he came to Holloman. The project was classified, and initially not even Dittmer knew why he was training the chimps. In an interview with George House, then curator at the New Mexico Museum of Space History, Dittmer said, "I never questioned anything because a lot of things at that time were classified and the less you knew of classified, the easier it is to keep it classified." But it soon became clear to him that the air force planned to use these chimps in rocket research. Dittmer was inducted into the International Space Hall of Fame in 2001 and died at the age of ninety-six in 2015.

As with the Soviet team working with dogs, the care of the chimp colony was of great importance to Colonel Stapp, the ranking officer on the project. "Stapp had chimpanzees for years prior to this—and he was very, very particular with the colony," Dittmer said. "Anyone that mistreated an animal or anything else, they were out the door. He didn't put up with any nonsense with his animals." The veterinary staff too were "all very professional people," Dittmer said, "and they didn't want no monkey business as far as the animals."

Among the staff working on the project, it was Dittmer who was closest to Ham. He had daily contact with Ham, preparing him for his training, guiding him through that training, and overseeing every aspect of his care and well-being. Dittmer's relationship with Ham (and Enos too) went beyond that of re-searcher/subject. Ham was a kind of colleague, a chimp with a job to do who relied primarily on Dittmer to help him do it. To do that job, they had to establish a strong working relationship. "I had a very good relationship with Ham, I think," Dittmer said. "I think—I know he liked me. I'd hold him and he was just like a little kid. He'd put his arm around me and he'd play, you know. He was a well-tempered chimp."

On January 31, 1961, Dittmer prepared Ham for his flight into space. The team had moved from Alamogordo to Cape Canaveral, Florida, to acclimate weeks ahead of the flight. Training went on as usual. Six chimps were candidates for this first test flight, and Ham was not officially chosen until the day before liftoff. His second was a chimp called Minnie, who would also be prepped alongside him.

Ham would fly on a Mercury-Redstone 2, a liquid oxygen–fueled rocket modified from the Redstone ballistic missile developed by Werner von Braun and his US Army team, the same basic design that put the first US satellite into orbit. Ham would not be flying into orbit, however, but out over the Atlantic where, after reaching an altitude of 115 miles, the capsule would separate from the rocket body, descend on braking chutes, and land in the ocean. Eight navy ships waited in the drop zone to pull the capsule from the water. During the flight, Ham was expected to reach a top speed of 4,400mph and endure about 9g. At 37 pounds, Ham would feel like he weighed 333 pounds.

Dittmer attached electrodes to Ham to monitor his heart during the flight and a respiration sensor. He then helped Ham into his flight suit, strapped him into his capsule seat, attached the shock plates to his feet, and sealed the cover. At 6 a.m. Dittmer accompanied Ham to the rocket. At 7:10 a.m.

Dittmer made a final check, looking in at Ham through the window in the cover of his capsule. "It looked like he was smiling at me," he said in his interview with House.

With Ham ready to fly, the rocket sat on the pad, waiting. Liftoff was scheduled for 9:30 a.m., but twice a team of engineers had to make repairs to bring down the rising temperature in Ham's capsule. As the team made their way down from the tower the second time, the elevator malfunctioned, and that too had to be repaired. It wasn't until 11 a.m. that the countdown resumed, and now the weather became a concern, as out over the Atlantic a line of storm clouds was building. It isn't the wind and rain that pose a danger to a rocket, but rather the possibility of lightning. A lightning strike can damage a rocket's guidance system and payload, and because rocket fuels are so flammable the whole thing can blow up. The team waited until they got the all clear, and the rocket finally lifted off a few minutes before noon. By this time Ham had been strapped into his capsule for some six hours.

In flight, more problems developed. The engines were burning fuel too fast, resulting in the rocket traveling too fast too soon. Its angle of ascent was sharper and steeper than planned, which meant the recovery teams had to adjust for a new splashdown location. The fuel ran out several seconds early, triggering an abort sequence, which fired and separated the spacecraft from the booster with the force of 17g, a crushing pressure that held Ham's limbs and head pinned against his seat, pressed back the flesh of his face, and must have made it nearly impossible to breathe. Ham lost focus and abandoned his task at the control panel. The force against him eased up and the electrical plates beneath his feet kicked in, jolting him back to his purpose.

Black-and-white file footage taken by a camera mounted in the *Mercury* spacecraft is grainy, and it stops and starts in shifting frames like an old-time silent movie, but you can see Ham during his flight, the camera catching the top of his head in motion, and occasionally a portion of his eyes and face. At times his head bobs about in a quiet rhythm, as if he's motoring down a dirt road in a pickup truck, and then it tosses violently forward and back, his eyes closed, his mouth opened, his lips stretched back, his white teeth and sharp canines visible as he bears the acceleration of the rocket. And then he returns to that soft bobbing as the spacecraft enters microgravity and his eyes come into view in the film, calm and dark. He appears to be completely at ease, performing his tasks at the control panel.

While Ham performed his tasks, the ground crew accounted for the anomaly during engine burn. Ham would remain in microgravity for one minute longer than planned, because instead of 115 miles altitude the rocket had achieved 155 miles altitude and would splash down at least 130 miles farther down range. The anomaly also caused a sudden loss in pressure inside the spacecraft, a drop from 5.5psi to 1psi. Ham's capsule inside the spacecraft had its own pressure and air system, but had he been inside the *Mercury* without this protection, the low pressure would have killed him. It would have killed an astronaut too.

For six minutes Ham floated in microgravity performing his tasks, proving that when an American astronaut made the journey he too would be able to endure the acceleration of the rocket and work in space. Then the *Mercury* spacecraft turned back toward the Earth. On its descent, the braking chutes did not perform optimally, as part of the system had been jetti-soned early, so when the capsule hit the ocean's surface, it hit hard, and in rugged seas. The impact breached the hull and jammed a valve open, and Ham's ship started taking on water as it rode seven-foot swells. Of course, Ham didn't know his ship was taking on water, that it was going down, and probably wouldn't know even when it went down, when it went under, falling to the bottom of the sea, the air running out as he suffocated and died. He would feel a tightening in his chest, an increase in his pulse and respiration, perhaps an adrenalin dump in his body as he panicked, and then it would all slow down and he would drift into a forgotten sleep. But Ham still had time, and rescue was on the way.

It would take the closest ship, the USS *Donner*, nearly an hour to arrive at the scene. Navy frogmen dropped from a helicopter and attached tow cables to Ham's capsule. The pilot then drew back on the stick and pulled the spacecraft out of the sea. Suspended from the helicopter, some eight hundred pounds of water drained from the spacecraft as the pilot set it gently down on the deck of the *Donner*. Ham's flight, from liftoff to splashdown, had taken seventeen minutes.

The team removed the cover from Ham's capsule, and there sat the world's first space chimp, blinking. He was in good condition, though mildly dehydrated from his long wait on the launchpad, and he had bruised his nose during the hard landing. Ham was posed for photographs, one shaking the hand of the *Donner*'s commander, Richard Brackett. In the photo he appears calm, even happy, still strapped into his seat, his face illuminated by the sun. In another photograph Ham is being offered an apple. He is baring his teeth, as if smiling, as he reaches out for the apple with both hands, his body still strapped against his seat, his black eyes focused. It had been hours since he had anything to eat. That photograph circulated in newspapers all over the world. Here is Ham, the happy space chimp, the captions read, enjoying a postflight snack. He has just returned to Earth after a successful flight into space for the glory of America.

Some years later the renowned primatologist Jane Goodall examined these photographs of Ham. In an interview for the documentary film *One Small Step*, she remarked that the look on Ham's face "is the most extreme fear that [she had] ever seen on any chimpanzee."

▢

With the success of Ham's test flight, the US was now ready to send a man into space. They chose *Mercury* pilot Alan Shepard, but before they could get that done the Soviets put Yuri Gagarin into orbit. Shepard went up about a month later, and like Ham, he did not enter orbit. He went up to 116 miles altitude and came back down. Still playing catch-up to the Soviets, the US would next test its readiness to orbit a man around the Earth. That's where Enos came in.

Enos was not a friendly chimp like Ham. "He wasn't really mean" either, Dittmer said in his 2012 interview with George House. "He just didn't take to cuddling. That's why in any pictures you ever see of Enos, you don't see anybody holding him." When moving Enos from his living quarters to training stations, he had to be led by a strap connected to his wrist. He didn't care much for the company of people and was prone to nipping and biting when agitated, and the strap helped keep a bit of distance between the handler and those sharp teeth. When officials came to the base and toured the chimp colony, Enos sometimes threw feces at them as they peered in through the cage. And he had developed the unsavory habit, when he had such an audience, of pulling down his diaper and stroking his penis. Airmen at Holloman began to call him "Enos the Penis."

The rocket carrying Enos ascended into the sky on November 29, 1961, at 10:08 a.m. Enos endured a lift maximum of 7.8g, and then the rocket settled into its arc and pushed out into Earth orbit. The flight plan was to orbit Enos three times and bring him back down. During the flight and in orbit, Enos worked the levers on the control panel. According to Burgess and Dubbs, trainers had also incorporated rest periods in his work, and he took these rest periods as he was trained to do. At one point the capsule began to heat up, but then it returned to a tolerable temperature range. Following that, a wiring malfunction caused several undeserved electrical shocks through the bottom of Enos's feet, but he kept to his program, pulling levers and then resting as scheduled, the spacecraft whirling around the Earth.

Before flight, the medical team had fitted Enos with an internal balloon catheter to prevent him from playing with his penis. In training, they had first tried an external catheter, but

Enos quickly and easily removed it. Then they tried an internal catheter, but Enos pulled it out. So for his big flight, they inserted a balloon catheter, in which the catheter tube extends from a plastic balloon inside the bladder, inflated with water, the tube of which runs down through the urethra and out the penis end. To pull that out when the balloon was inflated would not only require a great deal of strength; it would also be very painful. Enos left it alone, but certainly he didn't like it. And why did the team go to such trouble to prevent Enos from masturbating anyway? Was there some danger to masturbating in space? Would it interfere with his tasks onboard the spacecraft, or was it rather that the ground crew found it unsightly?

In flight, Enos worked the control panel, pulling this lever and then that lever in sequence, as he was trained to do. He did everything just right, but the wiring malfunction kept the shocks coming into the bottom of his feet. Zap. Zap. Zap. He worked the levers faster to get those shocks to stop. They always stopped, those shocks, when he worked the levers properly. And he was working the levers properly. It was the only thing he knew how to do, but the shocks kept coming. Zap. Zap. Zap. He hammered away at the panel. Zap. Zap. Zap. He hammered away. And then, his frustration rising, he came into awareness of another awful problem: what was this thing down his shorts? Enos reached into the folds of his flight suit, grasped the lead end of the bubble catheter, and pulled it out. The pain must have been unsurvivable, drawn deep from the middle of his body as the bubble, still inflated with water and the long tube attached, came sliding out. But then it was over, and Enos was likely filled with a pleasant sensation of relief and ease. As that feeling of pleasure overtook him, Enos did the only sensible thing he could do: he reached down to dandle himself. We know this because it was all caught on film. What we will never know for certain is whether Enos was merely rubbing the source of his pain after pulling the catheter out or on his way to finishing off the world's first space jack.

Enos completed his first revolution of the Earth, and on the second trip around, somewhere over western Australia, the spacecraft began to tumble. The thrusters were not all working properly, and later the team would find that a stray piece of metal had clogged a fuel line. But in the moment they knew only that the spacecraft was tumbling, burning fuel erratically, and the problem was getting worse. If they let the ship make its third orbit, it might not have enough fuel to power the thrusters, stabilize the ship's attitude, and drop it out of orbit and back into Earth's atmosphere. With only twelve seconds remaining before Enos was committed to a third orbit, the team aborted the mission. They brought the *Mercury* spacecraft down, Enos still working the levers as he fell through the atmosphere, the spacecraft and capsule heating in its speed until it splashed down in the south Atlantic near Bermuda. The crew of the USS *Stormes* pulled it from the sea. When the hatch opened, Enos leaped into Dittmer's arms.

The flight had lasted 3 hours, 21 minutes, and Enos had lived and worked in microgravity for 181 minutes. The team discovered that the temperature inside his capsule had peaked at 106 degrees. Enos would not have been able to tolerate that kind of heat for a third trip around. Still the mission proved that the US was ready to put its first man into orbit, and on February 20, 1962, the great John Glenn achieved that feat.

What became of Enos, the chimp who tested the hardware to make Glenn's flight possible? He retired to the Holloman chimp colony, where about a year later he died of dysentery. As Burgess and Dubbs note in their research, there is no memorial to Enos, and after a routine necropsy his remains were probably thrown out.

◻

The year before American astronauts Neil Armstrong and Buzz Aldrin became the first humans to walk on the moon, the Soviet Union sent two Horsfield's tortoises around it. In their *Zond 5* spacecraft, the tortoises, along with some mealworms, wine flies, plants, seeds, and bacteria, were the first living things to make a circumlunar voyage. In the pilot seat rode a 154-pound mannequin with radiation detectors inside.

Soviet chief designer Sergei Korolev had been more interested in a crewed mission to Mars and was working on his gargantuan N-1 rocket to achieve that dream, but his government steered him to the moon. Why? Likely because in 1961 President Kennedy publicly announced America's commitment to putting a man on the moon by the end of the decade. The USSR had already racked up a string of firsts in the Space Race, but its leaders wanted to bag the moon first too. Like most everything the Soviets did in those days, their moon program was a state secret. They denied working on the project until 1990, when *glasnost* pulled back the curtain from a great many Soviet secrets.

Korolev was charged with leading two moon programs,

one to take a crewed spacecraft around the moon and back to Earth, and the other to land a crew on the surface of the moon. His death in 1966 was a major setback in those efforts, but his two teams kept on with their work. Trial after trial of Korolev's N-1 rocket resulted in explosion and catastrophe on the launchpad or just above it. By using a smaller rocket to launch the *Zond 5*, the circumlunar project was going along rather well. The US would not equal that flight for another three months with *Apollo 8*, which flew a crew of three astronauts around the moon.

The Horsfield's tortoise, sometimes called the Russian tortoise, makes a great aquarium pet because it is small (between five and ten inches long) and requires little food. The male Horsfield's is known for its wild courtship display, shucking and diving with his head and biting the front legs of the female, little tortoise kisses, to get her in the mood. If she accepts him, he will mount her from behind and sound a series of high-pitched barbaric squeaks.

On the night of September 14, 1968, the tortoises went up from Baikonur, the Soviet Union's massive spaceport in Kazakhstan. Temperatures were mild, not too hot and not too cold, and the vast and empty desert was lit by the afterglow of the cosmos. The rocket blasted off and rose into the starry sky above the desert, an upside-down candle ascending into the heavens. At the time of launch, the tortoises had already been in the spacecraft for twelve days with no food. Once in Earth orbit, the team parked the *Zond 5* for a while as they made a series of system checks, then the third-stage engine fired, and the spacecraft moved onward to the moon.

On September 18 the spacecraft rounded the moon, flying 1,200 miles above its surface, and headed straight back. It did not enter orbit. An attitude control sensor had failed on the flight out, and now on the return a second sensor failed, resulting in difficulty guiding the spacecraft as it reentered Earth's atmosphere. It would have to make what NASA called "a direct ballistic entry," like a bullet breaking through. The ground team would not know precisely where the capsule was going to land until it came very near to landing. It splashed down in the Indian Ocean on September 21. A Soviet Academy of Sciences ship, the *Borovichi*, located the capsule bobbing in heavy seas and recovered it while a US Navy patrol looked on. Just what were the Soviets up to? Oh, probably just beating the Americans in the Space Race again.

The capsule arrived in Moscow on October 7. It had been a good month now since the tortoises were sealed inside without food. When the Soviet team finally opened the capsule on October 11, they found that the tortoises had lost 10 percent of their body weight, but they were generally healthy and had powerful appetites.

◌

In 1972 five pocket mice flew to the moon on *Apollo 17*, the final moon mission. Because pocket mice are desert dwellers and do not require water (they take in all the water they need from their food), they make excellent space travelers. These mice, one female and four males—affectionately called Fe, Fi, Fo, Fum, and Phooey by the astronauts who flew that mission— were to test the effects of high-energy radiation on the body, especially on the retina of the eyes and on the brain and skin.

Each mouse had a radiation detector surgically implanted into its brain. You can imagine these devices like little hats on the top of their heads, transforming the mice into cyborgs.

High-energy radiation, or cosmic radiation, strikes the Earth without cease. These particles are mostly hydrogen nuclei traveling at near the speed of light. Earth's magnetic field and atmosphere slow these particles down, and they give their energy to them, so that we are perfectly safe down here on the surface. Life would not have evolved and flourished here without shielding from cosmic radiation. Outside this shield, these particles penetrate living tissue and damage or destroy it. As the Apollo moon missions were operating outside Earth's protective shielding, they were ideal for research on the effects of cosmic radiation on travelers from Earth.

Apollo 17 (and the other Apollo flights) consisted of a three-man crew: commander Eugene Cernan and pilot and geologist Harrison Schmitt would descend to the moon's surface, while pilot Ronald Evans would remain with the mice in the command module in orbit around the moon. The mice required nothing from the astronauts. They were set up in a sealed aluminum canister, inside of which were individual tubes, one for each mouse and one empty tube to help with ventilation. The tubes protected the mice from tumbling about too severely, but they were free to move inside the tube, where they feasted on a prepared seed mixture (about thirty grams per day each). An identical canister with five other mice would remain at NASA's Ames Research Center in California as a control study.

Apollo 17 launched at night from Cape Canaveral on December 7, 1972, the first night launch in the US carrying astronauts.

Half a million people turned out to watch as the massive Saturn V rose into the dark sky, lighting up the Indian and Banana Rivers. Designed and developed by Werner von Braun, Arthur Rudolph, and their team in Huntsville, the Saturn V is the largest, most powerful rocket ever to fly; it had to be to lift and deliver the hardware required for the moon missions. It stood taller than the Statue of Liberty, weighed 6.5 million pounds when fueled, and lifted 310,000 pounds of payload into Earth orbit. The payload alone is the equivalent of about thirty-one elephants, and pretty big ones too. Most everyone watching on-site and on television knew that Cernan, Schmitt, and Evans were on board, but few likely knew about Fe, Fi, Fo, Fum, and Phooey, those little four-legged beasts riding that great energy into the heavens.

Cernan and Schmitt spent three days on the moon in a region known as the Taurus-Littrow valley. They lived and worked out of the lander, *Challenger*, as if on a long weekend camping trip in a remote location with no atmosphere. Their primary mission was to sample lunar highland material (the lighter spots on the surface of the moon when you look at it from Earth) and investigate possible new volcanic activity (less than three billion years old). Each day they ventured out for about seven hours, driving their Lunar Roving Vehicle (which is still up there) to various locations to take measurements, collect rocks, and deploy explosive packages that, when detonated, would generate data useful in mapping the top few kilometers of the moon's crust. On the first day Cernan caught the hammer attached to his space suit on the right rear fender of the rover and broke it. He and Schmitt fashioned a new fender out of a lunar map, duct tape, and clamps from a telescope. The mission broke a number of records that still stand, including longest duration moon landing, longest duration in lunar orbit, and largest lunar sample returned to Earth.

What were the mice doing during all this time in the command module with Evans? Eating seeds and absorbing cosmic radiation. Crew members of the previous Apollo missions had reported seeing flashes of light when they closed their eyes, usually when they put the lights out in the spacecraft for a sleep period. These flashes, or streaks of light, occurred about once every thirty seconds. The flashes were not observed on the surface of the moon but during the journey to the moon, in orbit around it, and in orbit around the Earth. While the mice collected data with the radiation detectors (dosimeters) implanted into their heads, Evans wore a specialized helmet to track cosmic rays (the Apollo Light Flash Moving Emulsion Detector). The result was that, yes, the flashes were indeed caused by cosmic rays. The next question was, were these cosmic rays harmful, especially to the retina of the eyes and to the brain? When the mice returned to Earth, researchers might find out.

After splashdown on December 19, the mice were transported aboard the USS *Ticonderoga* to a medical facility in Pago Pago, the territorial capital of American Samoa. In a letter to Colin Burgess, co-author of *Animals in Space*, Delbert Philpott, the principal investigator for the pocket mice experiment, tells the story of transporting the mice to his lab. Philpott knew that in the moist tropical air of the Pacific islands the canister might heat up and kill Fe, Fi, Fo, Fum, and Phooey.

He would have to hurry to get them from the ship at dockside to the lab facility at Lyndon B. Johnson Hospital. Speed limits on the island were painfully slow and rigorously enforced, and in a moment of roguish genius, Philpott realized that the only vehicle on the island that could push beyond these limits was an ambulance. So off he went in an ambulance to pick up the canister. He found a note from one of the astronauts attached: "For what it's worth, I think I hear scratching on the inside." So they were still alive, but they wouldn't be for long if he tarried. The ambulance took off, breaking the speed limit to get the mice to the lab where they could crack open the canister and let them out, give them water and food and air-conditioning, whatever they needed.

Later Philpott got a phone call from the press. *What's wrong with the mice?* they wanted to know. They had heard the mice were transported by ambulance to a hospital. *Well, nothing in fact,* Philpott responded, *nothing out of the ordinary. The mice were always going to end up at the hospital, because that's where the lab is.*

But the press was not too far off in their concern, because when Philpott opened the canister, he found two of the mice in good condition, two others in a weakened state, and Phooey, little Phooey, was dead. But then, Phooey, Fe, Fi, Fo, and Fum were always going to end up dead, because following examination the four remaining mice were euthanized for further study. Their bodies were preserved and flown to the Ames Center in California, where researchers concluded that high-energy radiation did damage the retinas of the mice but only minimally. The spacecraft had protected them fairly well.

These findings gave Apollo astronauts a little peace of mind. In 2016, however, Florida State University physiologist Michael Delp and his team published a paper supporting findings that astronauts who flew in the Apollo missions outside the protection of Earth's magnetic field had an increased rate of cardiovascular disease mortality of four to five times that of astronauts who did not fly at all, or who flew only into low-Earth orbit. Mice were again used in some of this research. Cosmic rays, then, are a danger to astronauts (and to mice) and will remain a major challenge for future deep-space travelers.

☐

Skylab (1973–79), America's first space station, rose into orbit on a Saturn V rocket repurposed from the Apollo moon missions, the final mission of the Apollo hardware. It had been just twenty-five years since Laika became the Earth's first space traveler, and now humans and animals were living and working in space for weeks at a time. At the end of its life, Skylab fell out of orbit and burned up in the atmosphere. It was an international event. Some people feared it would rain debris down on top of them; others wanted it to, or at least wanted to find debris scattered in their backyards. Some people wore Skylab T-shirts with a bull's-eye as an attractant; others wore hard hats as a repellant. Skylab broke up somewhere over western Australia, and a seventeen-year-old named Stan found twenty-four pieces near his hometown of Esperance. Then the town fined NASA four hundred dollars for littering. But before all that Skylab was an orbital laboratory, the next step in America's space program after the beauty and power of Apollo. What made Skylab really sing was the NASA Skylab Student Experiment Competition wherein a panel of National Science Teacher Association judges selected science projects to carry into space, proposed by students from all over the US.

One of the best experiments was "Web Formation in Zero Gravity," proposed by seventeen-year-old Judith Miles of Lexington, Massachusetts. It was the first arachnid study in space and featured two female crowned orb-weaver spiders that became known as Arabella and Anita. The experiment tested motor response in the spiders to better understand how their central nervous systems operated in microgravity. The orb-weaver spider wants to spin webs, as without a web there is no food, and without food there is only death. So the orb-weaver spins webs, beautiful, geometrically balanced webs. If Arabella and Anita could spin such webs in space, it would indicate that their central nervous systems were operating normally. And if their nervous systems were operating normally, perhaps American astronauts' nervous systems were too. The experiment would also say something about the role of gravity in a spider's ability to spin webs and help determine whether spiders, and other Earth creatures, could adapt to life in space. In his analysis of the experiment, "Spider Web-building in Outer Space," published in the *Journal of Arachnology*, Peter Witt and his co-authors write that the fact that "spiders always run on the underside of a web or a bridge thread, hanging down as they move, makes one aware of the important role which the use of the animal's own weight plays in locomotion and silk production." What would happen to a spider's locomotion and silk production in an environment with limited gravity?

The orb-weaver is a hardy spider that can live for up to three weeks without food, as long as it gets plenty of water. It bears a distinct pattern of mottled white over its abdomen that looks a bit like a cross, hence it is sometimes called a cross spider. The female is larger than the male, and as these things sometimes go, she will eat him just after mating. She bites him, wraps him in her white silk, waits. When he is dead, she liquefies his body with her vomit and then consumes him. She will consume her web too, daily, the sticky part where her prey is caught and rendered, and then spin another to await the next good catch. She hangs head-down in the center of the web or hides in nearby foliage with one of her clawed legs resting on a signal line, and when that line jitters, she kills and feeds.

A crew of three men flew with Arabella and Anita: commander Alan Bean (who also flew on *Apollo 12*), science pilot Owen Garriott, and pilot Jack Lousma. Before launch, Arabella and Anita were fed houseflies (lucky for the men on board) and given a sponge saturated with water. They launched on July 28, 1973, for a fifty-nine-day mission.

On August 5 Arabella was released into a lighted box where she would have the space to spin her web. Back on Earth, another orb-weaver was released into the same kind of box as a control experiment. When Garriott opened the vial, Arabella would not come out. Garriott waited for a time and then gave the vial a good shake. Out came Arabella into the box, floating away in microgravity, her legs furiously pumping as she drifted, until she hit the wall of the box, reached out, and grabbed on. Garriott would record the events with video and still cameras.

During that first day, all Arabella could produce was a little punk of a web in the corner of the box. The next day, however, she finished it, but it was crudely constructed, inexact and unkempt, with lines diving off in various directions. The strands themselves were spun at various thicknesses, and were mostly thinner, weaker strands. On Earth, a spider can vary strand thickness to meet the needs of its own body weight, but the strands are usually uniform in the spinning of a single web. Here in space, without gravity, Arabella had trouble sensing how much she weighed, so in her confusion she attempted a number of different sized strands. The main point was that the web held together, and there she was, Arabella in her space web, the first in the history of the world.

On August 13 Garriott destroyed Arabella's web to see if she would remake it and if the new web might not be better constructed. But she didn't remake it. She hung on the side of the box doing nearly nothing. Garriott decided to feed and water both spiders—Arabella in the box, Anita still in her holding vial. He replenished the water in the sponges and then fed the spiders a bit of filet mignon, cooked rare, meat scraps from the astronauts' supper. Arabella extracted juices from the meat and then kicked the dried thing out of her web as she would a desiccated fly. Thus fortified, she went to work again, and this time she constructed a very nice, geometrically balanced web. It had taken her several days to adjust to microgravity, Garriott decided, but once she did, microgravity was no longer a deterrent to good web-building. In his evaluation of the experiment, Witt and his co-authors arrived at a similar conclusion: "There is a transition time during which spiders gradually acquire the skill to move 'competently' under weightless conditions."

After about three weeks of web spinning, Arabella was shuttled back into her vial, and Anita was released into the box. The results were pretty near the same. Anita's first efforts were feeble, and after a couple of days and a couple of tries she got the hang of it. She began to spin lovely little webs in space. "It is probably the absence of body weight which disturbed each of the two animals severely during the first days after release from the vial," Witt and his co-authors write. Then on September 16, Garriott found Anita dead in the box. He collected her body and returned it to the vial. She would return to Earth with the crew for further examination.

The crew splashed down in the Pacific on September 25, and now back on Earth they found Arabella dead too, likely from starvation. It was a long way to go for two spiders. During their flight, Arabella and Anita got a lot of press, their story and pictures in newspapers and on TV. They were famous spiders now, despite the fact that they were dead, and so their curled, dead bodies joined the permanent collection at the Smithsonian. The most interesting finding, write Witt and his co-authors, has little to do with observations about the nervous system but with "the ability of an invertebrate animal with as rigid a behavior pattern as orb-web construction which is relatively independent of experience to find alternate ways to complete a perfect trap for food and thereby increase its chance for survival."

◻

In 2007 the European Space Agency sent a community of tardigrades, also known as the water bear, into the vacuum of space on the *Foton-M3* mission. The water bear is a microscopic invertebrate animal left over from the Cretaceous, and likely even further back, which means it has been around for at least 250 million years. Under a microscope it does look like a little bear, with a barrel-like body, eight little bear legs with claws, and a snout-like mouth used to pierce the cells of the tiny plants, algae, and invertebrates that it eats. On Earth, the water bear is found almost everywhere. It flourishes in mosses and marshes where conditions are generally warm and moist. It has also been found in the most extreme environments: under crushing pressure at ocean depths, at dizzyingly low pressure at mountain heights, in roiling hot springs, and on barren ice shelves. It lives in the wet, warm tropics and in dry, sandy deserts. Nearly unkillable, it is one of the hardiest animals on Earth.

After ten days of exposure to space on the *Foton-M3*, the water bears returned to Earth alive. A human being exposed to space without protection has about 30 to 90 seconds to live, and up to 180 seconds for Chuck Norris. It's not just the absence of breathable air that is the problem. It's also temperature extremes, the absence of atmospheric pressure (vacuum), and radiation. The water bear withstood all of this for ten days. It managed temperature ranges from -272.8 degrees Celsius (three times colder than the coldest temperature ever recorded on Earth) to 151 degrees Celsius (hot enough to bake bread or slow roast a chicken). The vacuum of space had almost no negative effects on the water bear. And it withstood cosmic radiation at a level 1,000 times greater than what we experience here on Earth. The UV radiation from our sun is also over 1,000 times greater in space than here on Earth, and it fries cellular structures and DNA. At this intensity a number of water bears did die, but a number also survived.

How does the water bear do it? The water bear does it by dropping into a dormant state. It reduces the moisture content in its body to as low as 3 percent and hardens the membranes of its cells. In effect, it creates its own protective shell called a tun, a little spacecraft if you will, inside which it waits for conditions to improve. Caspar Henderson writes in his fantastic book *The Book of Barely Imagined Beings* that on Earth the water bear can remain in this state for 120 years. When conditions improve, the water bear rehydrates itself and goes about its business, a "micro, aqueous Phoenix," Henderson calls it, rising up from the dead.

There is something even more curious going on here too. In 2015 a team of scientists from the University of North Carolina at Chapel Hill announced that they had successfully sequenced the genome of the water bear. What they discovered is that upward of 17.5 percent of the water bear's genes are not water bear genes. They come from other species, from bacteria, plants, fungi, and various microbes. Until now, it was thought that such gene borrowing occurred at a much lesser rate among Earth species. The rotifer was the previous champ at 8 to 9 percent, while most animals will carry only about 1 percent foreign DNA. The water bear accomplishes its feat through horizontal gene transfer as opposed to sexual reproduction. As it is coming out of its dormant state and reconstituting its body with water, its cell membranes are porous, even leaky, allowing foreign DNA and other molecules to enter. Some of these foreign bodies may be of no help at all and may even cause death. But some of them work to the water bear's advantage. Natural selection takes care of the rest, and 250 million years later we have an animal that really is as tough as Superman. It may be that the water bear's ability to withstand extreme environments is based on the acquisition of these foreign DNA, just as the process of moving into and out of its dormant state to withstand these extreme environments allows the acquisition to occur.

Studying the water bear as it endures the vacuum of space may help answer questions about the origin of life on Earth. Some scientists see evidence for Mars and Earth sharing in the origin of life during the planet-forming age of our solar system, and it may be that life persists on Mars, deep underground—we do not yet know. In 2016 scientists working in Greenland found the oldest fossils yet known—communities of bacteria, known as stromatolites—and they were alive at a time when Earth and Mars were much the same. So if life emerged in the conditions on Earth at that time, it could have emerged on Mars too. Or it's possible that life began on Mars and then was transferred or migrated to Earth. Species on Earth have always traveled across great distances borne on the wind, flushed down great rivers, riding rafts of flotsam across the seas from continent to continent. So why not through space? Transpermia, as the theory is known, takes into account the possibility that life may travel through space from planet to planet, or even between solar systems. Perhaps life is in the business of traveling about our galaxy, traveling about the cosmos, and seeding suitable planets as it goes. Perhaps the water bear came to Earth from somewhere else. Perhaps it is not to the Earth alone that we belong, but to the great cosmos itself.

In 2011 the California-based Planetary Society, working with scientists in Russia and Germany, set up the Living Interplanetary Flight Experiment to test this theory of transpermia by sending the water bear (along with samples of various bacteria, eukaryotes, and archaea) to Phobos, one of the two moons of Mars. In fact, so likely was it that at least some of the samples would survive deep-space travel and arrive at Phobos alive that the International Committee Against Mars Sample Return positioned itself against the mission. The probability that the samples would contaminate Phobos, and possibly Mars, they said, was too high to risk the science at all. The mission flew anyway, but the spacecraft stalled out in Earth orbit. A programming error on the Russian rocket made it impossible to boost it out of orbit and on to Mars. It eventually burned up on reentry, with fragments crashing into the Pacific Ocean near Chile.

○

It is impossible to hear these stories of animals flung into space—whether they lived or died—and not feel something for them, to come face-to-face with the plight of these beings, some of them very much like us, some of them with whom we share our daily lives. It is impossible to hear how these animals were taken from city streets and out of jungles, where they had lived by their own will and interests, and not grieve for them. It is impossible to look at photographs of them, their bodies confined by little space suits, strapped into metal harnesses to keep them "safe," their faces bearing the g-force of rocket flight, or in a contortion of enforced anesthesia from which they never will awake, and not wonder how human beings could do such things to them. The photographs are immeasurably sad.

These stories and photographs return us to a central tension in this book, which is a central tension in us all. On one hand, we want to have what we want, go where we want, do what we want, at whatever cost; on the other hand, we love the living beings of the Earth, as we love ourselves, and we want that same care and love returned upon us. Caught in the middle of this tension, we suffer. And we suffer more, I think, to feel the suffering of another than we do even in suffering ourselves. I do not here speak of physical pain. It is not the body's suffering we cannot bear, not the physical death of these animals in the fiery crashes of rockets in flight, abandoned at sea in sinking capsules, expiring in the desert sun. What we cannot bear is the feeling of loneliness that rushes in when we hear their stories.

When we hear the stories of animals flung into space, we have to ask: Were they lonely when they died? Would I be lonely? In the end, I think, we want to know that we were not alone in life or in death, that our people, who traveled with us through our journey on Earth, will travel with us on the other side, our orbits each following the other to whatever end, whatever eternity. Somewhere inside us, we all understand that what we have is not ours, that this Earth, and all its beauties and darknesses, was not meant to last forever, that everything we've ever known will one day vanish without a trace. This kind of loneliness drains the world of color, drags time behind it like an anchor, and pushes the body into an unrecoverable lethargy where the very air—hot or cold—becomes unbearable, and the Earth seems a lifeless rock, a world of unbroken wastes and desolation. It is, for the lonely, as if the Earth were placed here, and we born upon it, just to ripen loneliness.

In their book *Animal Astronauts*, Clyde Bergwin and William Coleman write that a reporter once asked US Navy captain Ashton Graybiel how monkeys were selected and trained for their flights, and how these monkeys responded to the training. Did they train willingly, or were they forced? "These monkeys are almost volunteers," Graybiel said. "During the preflight testing, we didn't force a monkey to take a test if it objected to it." This is some consolation, but I think we have to accept that unless a monkey is doing what monkeys evolved to do, they probably don't want to be doing it. In the 1950s during the height of US rocket tests with monkeys, people from all over the world wrote to the air force volunteering themselves as replacements. People wanted to go up in those rockets so that the monkeys didn't have to. A few volunteered to help repay a debt they felt they owed, men serving time in prison, for example. Send me instead of a monkey, they were saying. The monkey deserves to have its life. Surely my life might be used for something more purposeful than sitting in a prison and spending taxpayer dollars. The air force declined.

THREE

○

The Making of a Space Dog

A central element of the human future lies far beyond the Earth.

CARL SAGAN
Pale Blue Dot, 1994

In fall 1957, under the leadership of Colonel Sergei Korolev, an engineer in the Red Army, a secret team of Soviet rocket scientists and engineers went to work on *Sputnik II.* With this second satellite, Khrushchev wanted something new, something special, something that would again demand the world's attention. Drawing from the space dog training and flight program already in place, Korolev suggested putting a dog into orbit. The anniversary of the Bolshevik Revolution was only a month away, and after the success of *Sputnik I* the team had been released for a needed vacation. Korolev recalled them immediately to make good on Khrushchev's order. The team did not have time to develop a system to bring the satellite and the dog back safely. Their order was just to get the satellite up and beat the Americans again.

While this is the way the story unfolds in most accounts,

Sergei Khrushchev maintains that it was not his father pushing for the hasty launch of *Sputnik II* but Korolev. "Historians
in the rocket field ascribe [the launch deadline for *Sputnik II*]
to Father, and even say that he ordered it," Khrushchev writes
in his biography of his father, *Nikita Khrushchev and the Creation
of a Superpower.* "I find that very unlikely. Father understood
that as far as technical matters were concerned, he was not in
charge. Everything depended on the chief designer, and not
even on him, but on the degree of readiness for launch.... My
sense is that Father asked Sergei Pavlovich [Korolev] whether
it would be possible to schedule another launch to brighten
the holiday, and Korolyov quickly seized on the remark."

Here at the dawn of the Space Age, Korolev was approaching the height of his power and influence, but in the early
part of his life he had spent nearly seven years in prison. As
a young rocket and aviation engineer and project manager,
Korolev was interested in liquid-fueled winged vehicles, while
his colleague and rival, Valentin Glushko, wanted to focus on
solid-fueled rockets. Piqued by this difference, Glushko and
two other colleagues (who had also been arrested) denounced
Korolev as a subversive, holding up progress during that period of political repression under Soviet leader Joseph Stalin
known as the Great Purge (or sometimes the Great Terror).
Paranoid and ruthless, Stalin's government murdered hundreds of thousands of Soviet citizens, with some estimates
putting that number over one million. Millions more died in
gulags under forced labor and due to impossible living conditions. It is tempting to regard Glushko as the villain in this

story, but as many historians point out, Stalin made everyone the enemy of everyone else.

Stalin's secret police arrested Korolev early in the morning on June 27, 1938, at his home in Leningrad (now Saint Petersburg). His wife stood by helplessly as he was taken away and shipped off in a boxcar, unable to say goodbye to his three-year-old daughter, Natasha, asleep in the next room. During his imprisonment, Korolev's wife divorced him.

Tortured into a false confession and sentenced to ten years of imprisonment, Korolev was transferred from prison to prison, and then in 1939 he was sent to Kolyma, a gulag in far eastern Siberia notorious for deplorable living conditions, hard labor, and torture. In *Kolyma: The Arctic Death Camps*, Robert Conquest writes that conditions were so bad that more than two million people died at Kolyma alone. By the time Korolev's nightmare ended, he had developed a heart condition, suffered a broken jaw, and lost all his teeth. He spent the remaining years of his imprisonment in *sharagas*, prisons for intellectuals, especially engineers and scientists. He was discharged in 1944, but his charges were not officially dropped until 1957, the year *Sputnik I* and *II* went up. In 1945 Korolev was given the Badge of Honor and then commissioned a colonel in the Red Army. Later, when news surfaced that the US planned to launch an artificial Earth satellite as part of the IGY, Korolev convinced the Soviet Academy of Sciences and the government that he could beat the Americans into orbit. And he did.

It is a wonder that men like Korolev devoted their lives to the government that condemned and imprisoned them. How do we explain this? Perhaps love of country and the hope for a better future are enough, but Korolev's greater love had always been his work: design and engineering, space exploration and travel, and the dream of traveling to distant worlds. He wanted to go to Mars. He was the kind of man who wrote his own story, who cleared obstacles from his path as he went, no matter the cost. He was fixated on a future he had determined for himself and for humankind. No government was going to get in the way of that. In his book *Korolev*, James Harford writes that in winter 1965 Korolev spoke with Josef Gitelson, a scientist who worked primarily on life support and ecological systems for the USSR's Institute of Biophysics. Korolev told Gitelson, "I have a short time before me, maybe ten years, and I want to send humans to the nearest planet." That nearest planet was Mars. "His original plan, even before concentrating on the Moon project," writes Harford, "was to launch a cosmonaut around Mars." He launched two uncrewed probes to Mars during his lifetime, but both failed.

Heavily built and muscular, with dark brown eyes and an unstoppable intelligence, Korolev's great talent was not so much design and engineering but project management, organization, leadership. He was a master collaborator, a man his team both respected and feared. He was fast to anger but lavishly generous. If Korolev told you to do something, you did it. In *Challenge to Apollo*, Asif Siddiqi reports that engineer Anatoliy Abramov wrote: "Korolev's diatribes were the stuff of legend, and he was a master at it; his eyes would flash, his words would destroy yours, he would threaten to send you home walking between the railway tracks." Despite such harshness, Korolev understood that in a system where people who worked hard earned little more than people who didn't, his team would benefit if he offered incentives by way of bonus money and elaborate vacations. "A real king," Anatoliy Kirillov calls him in *Roads to Space*, an oral history of the early days of the Soviet space program, "the undisputed boss of everything." That role as undisputed boss made him sometimes difficult to work with, while his prison years had left him soberingly pragmatic, even prone to dark moods. He was fond of proclaiming, "We will all vanish without a trace."

Like Laika in her capsule, Korolev was a kind of prisoner inside his work, as the Soviet government saw to it that he lived the life of a vanished man. His identity was known only to a select few in the government, to his family, and to the people he worked with. To most Soviets and to the outside world, he was known only as the chief designer. US intelligence did not know his name until a year before his death, and it was only after his death that the Soviets released his name to the world. Were his identity known, the Soviets feared, the US would assassinate him. So Korolev lived and worked his entire life in anonymity. Perhaps if he had been known to the world and lived in the light of his work, it would have helped him shrug off his great darkness. Perhaps not. But one thing the Soviets got right: genius is not rooted in the individual but in the community, not in the chief designer alone but in his team. It is a collaboration of great minds, each doing their part, nurtured and driven by competition. Korolev's genius was in the way he orchestrated all the parts to make them work together. In his mind was a future he wanted to build, and he found the will and the people to help him build it. Among the great achievements of Korolev and his team is a string of world firsts: the first artificial Earth satellite (*Sputnik I*, 1957); the first biological satellite (*Sputnik II*, 1957); the first moon landing, uncrewed (*Luna 2*, 1959); the first human being in space (Yuri Gagarin, 1961); and the first spacewalk (Alexey Leonov, 1965).

Korolev's genius is an extension of the work of earlier and contemporary scientists and engineers who informed and inspired him. First, take Russia's Nikolai Kibalchich (1853–1881), who was hanged for building the explosives used to assassinate Tsar Alexander II in 1881. Imprisoned and awaiting execution, Kibalchich wrote down what he called his grand idea, essentially that explosives could be used not only to kill a tsar but also to propel a flying machine through the atmosphere. Soviet science writer Yakov Perelman later described his grand idea as "the first step in the history of spaceflight." Konstantin Tsiolkovsky (1857–1935), a high school math and physics teacher, worked on designs for a spacecraft capable of taking men to distant planets, namely Mars. He predicted that satellites and crewed space stations would one day orbit the Earth, and that this might be achieved by use of a rocket burning liquid fuels. He is best known for the Tsiolkovsky equation, his calculation for the speed required for a spacecraft to achieve Earth orbit—five miles per second, or about 18,000mph—and that this speed could be achieved by a multistage rocket fueled by burning liquid hydrogen and oxygen. These same basic principles govern rocketry today. "There is no doubt, though," writes Harford, "about the degree to which Tsiolkovsky's works gave direction to Sergei Korolev. In simple fact, Korolev began to build what Tsiolkovsky had conceived." Yet for Tsiolkovsky, what he had conceived was a mechanical means to a much greater achievement: happiness for all beings in the universe. In order to achieve happiness (what Tsiolkovsky defined as the absence of suffering), humans had to understand the workings of the universe, which was only possible by learning to live and work in space. It was inevitable, he believed, that human beings would one day live on other planets throughout our solar system and eventually spread across the galaxy. For Tsiolkovsky, our technology, our will and choice, our migration out among the stars, would one day drive our evolution, and we would then be able to remake ourselves into something new, something better.

From the US, Korolev was inspired by the work of Robert Goddard (1882–1945), who built and launched the world's first liquid-fueled rocket in 1926, apparently from the farm of his aunt Effie in his hometown of Auburn, Massachusetts. He later moved his work to Roswell, New Mexico. Among his achievements, Goddard launched 35 rockets, filed 214 patents, developed the first gyroscopic stabilization systems, steerable thrust, and parachute recovery systems, and wrote one of the still classic books on rocketry, *A Method of Reaching Extreme Altitudes*. Goddard kept a series of notes and musings locked away in a friend's safe that describe what he calls "the ultimate migration," a human voyage to distant worlds in ships propelled by atomic energy or a combination of hydrogen, oxygen, and solar energy. Another option Goddard toyed with was to hitch a ride on a comet, an idea later put forth by NASA as a viable option for deep-space travel. He also imagined that human beings might be able to withstand long space voyages by dropping into a cryogenic state.

And, finally, among this company is Werner von Braun from Germany (1912–1977), who came to the United States after designing and building the V-2 missile for Hitler during World War II. After the war von Braun went to work for the US Army and later for NASA, turning his great talent and ambition away from missiles for military application and toward exploration of the cosmos. He was the chief architect of the Apollo program's Saturn V rocket, which took human beings to the moon. Like many of his predecessors and contemporaries, von Braun argued passionately for sending human beings to Mars.

In an interview with Sergei Khrushchev, I asked about Korolev, a man he had met socially through his father. "Korolev was a tank," Khrushchev told me. "A brilliant manager, not really a scientist, but he was the person who could push everything through. And he could organize everything around him and keep control, like a general. He was not a general like Eisenhower, maybe more like MacArthur. Or maybe better to say General Patton. He was a very good man. It would have been very difficult to bring these projects to success without his ability to organize all these people."

☐

In 1949 Korolev selected a young physician named Vladimir Yazdovsky (1913–1999) to lead a team in biomedical research to prepare the way for sending a human being into space. Yazdovsky was charged not only with designing and building life-support systems for space travel but also with developing a selection and training program for small animals that would test those systems. After working as a physician for the army during World War II, Yazdovsky had hoped to retire from the military, but when his request was denied he continued his service with the air force, working on aviation medicine. When Korolev approached Yazdovsky with an offer to work on a new project, he was initially uninterested. "I informed [Korolev] that…I was already committed to other work in aviation medicine," Yazdovsky said in *Roads to Space*. Korolev, naturally, pushed back. "What I am offering you is far more challenging," he said. Then he asked, "Have you ever watched a rocket being launched?" Yazdovsky responded that he had not. "Well, then," Korolev said, "if you've seen it once, it will stay with you for the rest of your life."

As a guest of Lockheed Martin and United Launch Alliance, I watched an ISS resupply mission launch on an Atlas V rocket out of Cape Canaveral in spring 2016. The day before the launch, I joined a party of other guests—mostly family and friends of the teams that built the rocket—to watch the rocket rollout to the pad. It moved slowly along a track, like the Empire State Building on wheels. The next evening, after hors d'oeuvres, we assembled on the roof of a building some four miles from the launchpad, which was the closest we could get without being part of the team launching the rocket, a United Launch Alliance representative told me. A few clouds drifted in moonlight over the Atlantic, but it was otherwise a clear and starry night. I stood near a couple of people I had met there, the man who directed the team that designed and built the rocket's wiring harness, and an executive with a top NASA contractor. People drew in their breath when the engines fired, and the rocket came up into the space above the pad and seemed to hover there. When the rocket rose against the night sky, the bright flare from the engines roared, and then it crackled, like violently crumpling brown paper. That sound filled everything that could hold a sound and then spilled over. I could feel it in my feet and in my chest. I felt something of myself going up with that rocket too, that in my brief time in its company I had become somehow invested. It moved out and away, and I watched as the sky closed beneath it. "God," I heard someone say. "I bet that thing gets terrible gas mileage."

I'm pretty sure seeing that rocket launch will stay with me for the rest of my life, as Korolev had said, but that claim still wasn't enough to convince Yazdovsky. He remained committed to his current obligations, so Korolev took a number of evening walks with him in Moscow's Petrovsko-Razumovskiy Park to talk him through the details. "He explained everything to me," Yazdovsky said. "Then he took me to his design bureau, showed me around at the plant where rockets were being manufactured, and introduced me to his colleagues." Korolev believed in pushing hard at the limits of things; pushing hard was how he lived his life. He told Yazdovsky that life wasn't really life without risk, and that there was nothing more beautiful in the world than watching a rocket ascend into the sky. Korolev's patient insistence paid off when Yazdovsky finally accepted the position. As a first step, Korolev directed Yazdovsky to speak with the man who was currently working on the project and could catch him up on what had been done so far. "I did as I had been told," Yazdovsky said, "but found nothing except for a sheet of paper with an elaborate sketch of a dog drawn on it. I told Korolev that we would have to start from scratch."

Careful, meticulous, and unswerving, Yazdovsky went to work developing a space dog training program. By the close of his long career, he had racked up an impressive list of honors and awards, among them the unofficial titles of "pioneer" and "founder" of Soviet space medicine and biology. It was Yazdovsky who selected and prepared the space dogs to fly, and it was Yazdovsky who carried the burden of responsibility when those dogs did not make it back.

The decision to work with dogs in the Soviet space program, as opposed to some other small animal (monkeys, rats, or even cats), grew out of Russia's history of working with dogs in scientific research, made famous by physiologist Ivan Pavlov (1849–1936) and his work in respondent conditioning. Biological rockets in the United States were carrying monkeys, and monkeys proved difficult to train and vulnerable to the stresses of spaceflight. They just weren't very tough. Yazdovsky knew this because he had read books in translation written by Americans working on these flights. The books "proved very helpful," Yazdovsky said, "since I could well appreciate what the Americans had been up against." The Soviets knew dogs, and they knew how to work with them. Dogs are easy to train; akin to humans in their physiology and their emotional and physical reactions to stimuli; easy to care for; and easy and inexpensive to acquire. In order to put a man into space,

Yazdovsky's team would have to first study animals subjected to the conditions of spaceflight, and they were going to do it with dogs.

Yazdovsky went to work identifying the major challenges of spaceflight. It would be a tedious process, taking small steps only, inching ever forward. With each problem solved, new problems were uncovered. This is the way of science, the way of new technologies. "The machine," writes Antoine de Saint-Exupéry in *Wind, Sand, and Stars*, "does not isolate man from the great problems of nature but plunges him more deeply into them." If the Soviets were to put a man into space and do it safely, Yazdovsky determined, they would need to address three major challenges:

1. the conditions of near-Earth orbit, which includes the absence of oxygen, meteorites that might damage a spacecraft and its crew, cosmic and solar radiation, and extreme temperatures both hot and cold

2. spaceflight itself, which includes acceleration and the resulting g-forces, vibration, weightlessness, and noise

3. confinement in a small spacecraft, which includes isolation; bodily functions, especially eating, drinking, and elimination of waste; and psychological stresses

Soviet rockets in the early stages of development could not carry the heavy payloads of today, so space dogs had to be small, between thirteen and sixteen pounds. They had to be relatively young, between eighteen months and six years. And they had to be white, mostly white, or another suitably light color. A black-and-white video camera was installed in many of the spacecraft to gather visual data during flight, and white dogs were much easier to see in this footage. Finally, the dogs had to be female. A primary reason for this was waste collection. Flight suits were fitted with a waste collection system that functioned best when both solid and liquid waste exited the dog from the same basic location of the body. There was no equipment or room in a flight capsule allowing a male dog to lift his leg. Even so, a number of the sources I read refer to some of the space dogs as "him" or "he," and some were given names more suitable for male dogs. A Russian friend and translator explained to me that in Russian one would always refer to a dog with a male name as "he," regardless of its sex. So those dogs were female with male names, or some were in fact male. Indeed, a few sources report that a few male dogs did fly but only on suborbital flights. Other sources avoid using pronouns for the dogs altogether, preferring their names, possibly to avoid inaccuracies regarding sex.

When evaluating temperament, the dogs were sorted into three categories: even-tempered, anxious or restless, and inactive. Even-tempered dogs were generally the most successful in the training program, but a restless dog might become more even-tempered, or an inactive dog might come alive when given a job to do and a routine to do it in. The dogs eventually identified as suitable for flight were later sorted into rocket dogs qualified for short-duration suborbital flight and satellite dogs qualified for long-duration orbital flight.

Most of these small white female dogs were acquired at shelters in Moscow or directly from the city streets. "I went around with a tape measure and some tasty morsels," said biologist Ludmilla Radkevich in the documentary *Space Dogs*. "For several days I drove around in a military car with a military driver through the outskirts of Moscow. If we saw a little dog running along, I'd jump out, measure it, and if it was small enough, pop it in the car. We collected about sixty dogs that way."

The team preferred mixed-breed dogs, primarily for their hardiness. "We did not choose purebred dogs for the flights as they lacked the required resilience to the flight loads," Yazdovsky said in *Space Dogs*. "Instead we picked mongrels which were more accustomed to extreme living conditions." If a dog was alive and well on Moscow's city streets, the thinking went, it could endure cold and other challenging weather conditions and isolation, and go without food for long periods of time. The streets of Moscow turned out to be the perfect environment for the making of a space dog.

The dogs' health was primary to the success of their training and flight program and to the science, so important that Korolev himself made daily visits to the kennels to check on the dogs and on their caregivers' efficiency and attention. He did not tolerate sloppy work. He loved the dogs and often picked them up and whirled them around in his arms. It was often overly warm inside the kennels, and the soldiers guarding the dogs were tasked with keeping their water bowls filled. As Korolev made his rounds one day, he found a few of the bowls empty. "Let's get someone in here who cares about dogs," he reportedly said, and sent the offending soldier to the brig. "It's hard to believe that someone could love our dogs so much," medical doctor Alexander Seryapin said in *Space Dogs*. "First thing every morning, [Korolev] went not to his office but to visit the dogs. He would check the temperature of their water. If the water was warm, he would give the lab technicians hell. 'Why is their water warm? What have they been fed today?' And every day after work without fail, [Korolev] dropped in to see how the dogs were doing."

Once selected, training helped the dogs cope with the conditions of flight. While the dogs were not volunteers, a great deal of care was taken to put only the dogs that completed the training into rockets. The dogs trained to endure confinement for long periods of time (up to twenty days for long-duration flights), the noise and vibration of rocket flight, and a high-g environment followed by microgravity. The noise and vibration could be simulated by a vibration table. The dogs in training were secured to this table, Burgess and Dubbs write, and "the hapless animal would be shaken about, while the mechanism of the table created a loud and obviously frightening banging." Sensors recorded blood pressure and heart rate, which rose dramatically during the vibration and noise but then returned to normal after the test. The g-force of rocket flight was simulated in a centrifuge, which spun the dogs in a circle on the end of a mechanical arm. While the acceleration of liquid-fueled rockets tops out at about 5g, the training pushed the dogs to a maximum of 10g. In a high-g environment, blood moves away from the brain, which can result in a loss of vision and consciousness. Even today, pilots and astronauts train in such centrifuges to build tolerance and familiarity with high g-force.

☐

Sputnik II was a cone-shaped satellite sitting on top of a modified R-7 rocket, the world's first ICBM. The satellite consisted of three main components, stacked one on top of the other. On top was a small cylindrical instrument package with two spectrophotometers for measuring solar radiation, especially in the ultraviolet and X-ray regions of the spectrum, as well as cosmic radiation. In the middle was a spherical container that had been built as a backup for *Sputnik I*, containing batteries, radio transmitters, and other instruments. And on the bottom, Laika, inside a cylindrical pressurized capsule with its small round window. When completed, *Sputnik II* measured 13 feet long and 7 feet wide at its base and weighed 1,120 pounds, six times heavier than *Sputnik I*.

Laika's capsule was made of aluminum alloy and measured 31.5 inches long by 25 inches in diameter, not very big at all, but then she wasn't a very big dog. In his book *Korolev*, James Harford remarks that the capsule was not built specifically for Laika; rather, the team modified the capsule that housed Dezik and Tsygan on the first suborbital space dog flight in 1951. Inside the capsule Laika was positioned on a cork floor with sides coming up above her, a slot, really, just her size. Restraints limited her movement and secured her during the rocket's ascent and in orbit, and helped ensure that she did not tear loose from the sensors attached to her body. Despite these restrictions, Laika could lie down comfortably, sit hunched over, and move forward and backward, but she could not turn around.

The small round window made the capsule look as if it had an eye, and it was sometimes called a "Cyclops chamber." A crate or kennel used to transport or confine dogs is similarly designed, and when inside, a dog will almost always face the door, the end where the most light enters, not unlike a wolf in its den. It makes sense that a denning animal like a dog would fare much better on an extended spaceflight than a monkey or a chimp evolved to live in forests and forest canopies and on open scrublands. These considerations must have driven the design and installation of the window, but the primary purpose had to be observation from the outside. The caregivers and scientists used the window to monitor Laika during training and preflight preparations. During flight, the window was of use only to the dog. Some sources speak of a video camera mounted inside Laika's capsule, but they are surely in error. In all my searching, I could find no photographs or video footage of Laika, either during liftoff or in orbit. The ground crew had no way to visually monitor her after they secured the protective fairing over the nose cone of the rocket.

One of the problems the team struggled to solve was carrying enough battery power for the duration of Laika's mission, which was planned for seven days in orbit. Today's network of satellites and ground stations makes it possible to transmit and receive information nearly anywhere on Earth. In 1957, however, *Sputnik II* had to fly over a ground station that could receive its signal for only about fifteen minutes, after which it passed out of range. Why keep putting the signal out if there was no Soviet station below to receive it, the team asked? To save critical battery power, they installed an automatic switch that turned the telemetry system on only when the satellite was passing over such a ground station.

Simply put, a rocket is a payload sitting on top of a fuel tank with an engine. The majority of its weight comes from its fuel, not its payload, but the payload has to be within range of what the rocket can lift into space beyond its own weight. Until *Sputnik II*, nothing over 180 pounds had ever been lifted into orbit, and that was *Sputnik I*. Of primary importance then was keeping *Sputnik II* light, slimming it down wherever and however possible. The team decided it would keep the final stage rocket booster attached to the satellite, allowing the elimination of the pyrotechnics and hardware required to separate the booster. Some sources report that the booster failed to separate in space, but this is not true. Another reason to keep the booster attached to the satellite was to use its telemetry system to send the data collected from Laika's sensors to Earth. The telemetry system on the satellite itself was maxed out transmitting two different signals to Earth: a steady tone at 40MHz and a pulsing tone at 20MHz that was identical to that of *Sputnik I*. There was no more room in the satellite for a second telemetry system for Laika's data, so the team had to use the system in the booster. Finally, the team had learned from *Sputnik I* that the booster was much brighter in orbit than the satellite and so possible to see from the ground. They would leave the booster attached to *Sputnik II* because, as a November 13, 1957, article in *Pravda* reports, it "appreciably simplified the task of ascertaining the sputnik's bearings by means of optical observation."

What was unclear, however, was how the attached booster

would affect the temperature inside Laika's capsule. The team knew the sun would heat Laika's capsule from the outside, and the instruments and batteries, as well as Laika herself, would heat it from the inside. To cool Laika's capsule, the team relied on a forced-air cooling system, insulation, and the satellite's metal exterior, which would help reflect heat back into space. But the system could not be tested properly on the ground. The only way to know if the system worked was to launch the satellite, with Laika inside, into orbit.

◌

Sometime before the revolution of agriculture, about 15,000 years before the present, or according to some scholars, even deeper into our ancestral past, 30,000 years, maybe even 100,000 years, the wolf and the human animal came together. Some say it happened first with the Asiatic wolf in present-day China. Others assert it happened among the Natufians of the Upper Jordan Valley in present-day Israel. Or perhaps it happened in several places and times when the human world was inseparable from the world of animals. Wherever and whenever it happened, the wolf and the human animal came together, and we have not been separate since.

The process of domestication is a human enterprise, something human beings do to make use of animals for work and food. "The dog is an animal domesticated," writes Alexandra Horowitz in *Inside of a Dog*, "a word that grew from a root form meaning 'belonging to the house.' Dogs are animals who belong around houses," and it was humans who brought them around and brought them in. But domestication is something animals have done to themselves too. We made dogs central to our lives, just as dogs made us central to theirs. "The dog is a member of a human social group," writes Horowitz, "its natural environment, among people and other dogs." This has been, as many scholars have noted, a positive arrangement for dogs, which have proliferated in every climate and on every continent, while wolf populations are declining worldwide.

You can imagine the conditions under which the dog came into the house, a world in which a few less aggressive wolves drew close to human camps and picked up scrap meat and bones. Their pups were likewise conditioned to tolerate the presence of humans, and little by little wolves moved into the camps as the people allowed. Tolerant wolves like these living in and around camps with humans would have defended their food source from more aggressive wolves and other animals. At some point humans must have come to understand that these wolves were helpful and, then later, essential to daily life. Over time these wolves became dogs.

Work in exchange for food, maybe shelter too, is the original arrangement between humans and wolves, and when wolves became dogs this agreement remained intact. Over the past 15,000 years, dogs have also been food for humans about as often as they have been fed by us. Only the wealthy and noble classes had the kind of resources to keep dogs as pets, that is, until recently with the rise of a global middle class. Most people throughout human history could afford to keep a dog only for food or work. In making our living, we have used dogs as draught animals, capable of pulling a small cart, sled, or travois; for herding; for guarding property and people; for service in war and police operations; for hunting; for search and rescue; and more recently for therapy and service to people with disabilities. Dogs have performed so many wondrous tasks. In the Middle Ages and for a time after, a breed of short-legged dog was trained to run on a wheel like a hamster. The energy produced by the wheel could be harnessed to turn meat on a spit or churn butter. Such dogs were known as turnspit dogs and were the cornerstone of the modern medieval kitchen.

◻

In preparation for what she would experience inside *Sputnik II*, Laika learned to endure confinement for long periods of time. She did not achieve this all at once but rather in a series of small steps. Wearing a restraining suit with attachment rings, she was put into a small capsule and metal restraining chains attached to the rings on her suit. She endured confinement this way for an hour before her trainers released her. The next training session she might spend two hours in the box. Then four. And so on. According to Burgess and Dubbs, the dogs "typically protested with barks and whining when placed in this restrictive space for periods of two to three days." Laika worked through this discomfort and settled down, which encouraged the team to advance her in the training. Eventually she was able to tolerate confinement in these capsules for up to twenty days. Like the capsule in which she would fly into space, these training capsules habituated Laika to a low-light environment. The only light inside came in through the window.

Inside her capsule, it would be impossible for Laika to get into a squatting position. She would have to do her best in those cramped quarters to eliminate into a waste catchment bag attached to her pelvis and secured by a shoulder harness. "By means of a rubber tube in the receptacle, the animal's excretions were drained into an airtight 'latrine' reservoir," write Chernov and Yakovlev in their 1959 report "Research on the Flight of a Living Creature in an Artificial Earth Satellite." "For the purpose of deodorization and the absorption of liquid fractions, the reservoir contained a certain amount of activated carbon and specially dried moss." This system was unnecessary for brief, suborbital flights but essential for longer duration flights. In training, most space dogs were reluctant to use the waste system. They just didn't bother to go, and even laxatives were little good in encouraging them. But slowly over time, some of the dogs learned to use the system, and one of those dogs was Laika. In some photos of Laika, you can see the catchment bag attached to her pelvis, hanging down beneath her.

If Laika was going to survive in orbit for seven days on board *Sputnik II* (the duration of her mission), she would need food and water. In microgravity, dry kibble and canned dog food would float around inside the capsule. "Igor Sergeevich Balakhovsky, who dealt with the problem of dog food in orbital flights, brilliantly solved [this problem]," writes Adil Ravgatovna Kotovskaya of the Institute of Biomedical Problems in her essay "Why Were Flying Dogs Needed for Rockets and Satellites to Launch Yuri Gagarin?" Balakhovsky developed a gelatinous mixture of food and water that would hold together and stick to a feeding tin. Laika and the other space dogs had to be trained to eat this food, because it was not very palatable. Such training probably consisted of introducing Laika to the food a little at a time and also withholding food until she was hungry enough to eat anything.

In the sources I consulted, space dog food is always referenced as providing adequate nutrition and water, but how could this be so? In their 1959 report, Chernov and Yakovlev write that the space dog food was made from "40 percent bread crumbs, 40 percent powdered meat, and 20 percent beef fat" mixed with water and agar to form a "gelatinous substance." A typical space dog, they report, required no more than 100 grams per day of the pressed food before it was mixed with water and agar. He goes on to say that the space dogs required between 120ml and 200ml of water per day. Presumably then, the formula for space dog food was a mixture of 100 grams of food with upwards of 200ml of water, plus the agar to hold it together. Two hundred milliliters of water weighs 200 grams, so the space dog food was mixed at a ratio of roughly one part food and two parts water.

Let's concede that 100 grams of pressed food (before the added water) was adequate for a space dog for one day. What about the water? How much water does a dog need in one day? According to veterinarians I've worked with, most dogs require a good deal of water, between 15 and 30ml of water per pound of body weight, per day. Laika weighed thirteen pounds, so she would need 195 to 390ml of water per day, depending on activity level and environmental conditions. Based on these numbers, Chernov and Yakovlev's space dog food meets only the barest minimum requirement for water. A dog can live for two or three days without any water, but by the third day, without emergency medical attention, the dog will surely die.

Instead of supplying adequate food and water, I think the space dog food was barely adequate. After a day or so eating only space dog food, a dog would be dehydrated, and without additional water it would continue on a downward spiral into death, probably within three days.

In *Roads to Space*, engineer Arkady Ostashov remarks that in a meeting the team discussed installing an automatic feeder in the capsule so that Laika could be fed daily to keep her in good health for as long as possible. But sensitive to reducing the weight of the satellite, Ostashov and other engineers suggested saving "a few kilograms by designing the feeder for one meal only, since [they] were mainly interested in knowing whether the dog would be able to eat at all." Though the medical staff led by Yazdovsky and Gazenko surely supported feeding Laika once each day, the team opted to follow the engineers' recommendations. They determined that Laika could survive for up to seven days without much loss of body weight on one feeding of a three-liter ration of space dog food. The food would be available to Laika when she was sealed inside the capsule before launch. What she did with that food was up to her. She could eat it all at once, slowly over time, or not at all. If she ate all the food before launch, the team would not be able to determine if she could eat in microgravity, so they were taking a gamble there. If she waited until after launch to eat the food, since they could not observe her from the ground, the only indication that she could eat in microgravity

catchment system for a dog that is going to die in space no matter what. Or perhaps you are developing food that is completely inadequate to keep a dog alive in space for more than a couple days. To save even more time, why not put the dog into the capsule and send it up with no waste catchment system, no food at all, and no water? Of course, the team was developing and testing systems for future human spaceflight, and the data collected from Laika on the performance of these systems was essential toward that end. But the point was to get that satellite up to celebrate the Bolshevik holiday and beat the Americans again. So why go to all the trouble of a waste system? Why go to all the trouble of developing food that wasn't going to keep Laika alive?

I think these efforts are a testament to the care and respect the scientists and engineers had for the space dogs. They did what they did for science, but also because it was ethical, because it was humane, because the space dogs were their partners and colleagues and deserved to be treated well, as well as possible under the circumstances. The space dogs were teaching them everything they needed to know to achieve human spaceflight, and the team recognized this great debt. "Like their American counterparts they were, in essence, writing the book as they went along—making things up, trying things out, pushing the boundaries of their understanding," write Burgess and Dubbs. Without the dogs, there would be no new book, there would be no new field of space medicine, there would be no pushing at the boundaries of understanding. The space dogs are the underpinnings of humanity's exploration of space, our partners and our companions in cosmic exploration. Without the dogs, there would be no spaceflight.

◻

In 1925 the Russian writer Mikhail Bulgakov wrote a curious novel titled *Heart of a Dog*. Better known for *The Master and Margarita*, a novel often cited as one of the great masterpieces of the twentieth century, Bulgakov had abandoned a career as a physician to focus on writing. Indeed, his main character in *Heart of a Dog*, Philip Philippovich, is a physician, a surgeon more truly, who rescues a stray dog from the Moscow streets by offering it a piece of sausage. The dog, whom he calls Sharik, has been severely burned by a cook who threw boiling water on him as he rooted through the garbage. Upon receiving the sausage, this pathetic, whimpering dog is so happy that he follows Philippovich home. Later Philippovich is asked how he was able to get such a nervous dog to follow him. "By kindness," he answers. "The only method possible in dealing with living creatures. By terror you cannot get anywhere with an animal, no matter what its stage of development."

As Sharik recovers from his wounds and settles into the good life, Philippovich reveals his sinister intent. He takes the dog into his medical lab, surgically implants a human pituitary gland in its skull, and then gives it human testicles. Sharik transforms into a man, but a poor example of a man, with the ill-beseeming behaviors of the low-life alcoholic from whom the organs were taken. As a result, Sharik turns Philippovich's household upside down. He leaves a spigot on and floods the apartment. He attempts to rape a woman, one of the servants. He takes a disgusting job strangling stray cats for the state, their furs to become coats for Soviet workers. All this, and some trouble with the law for his strange experiment, prompts Philippovich to reverse the surgery, transforming Sharik back into a dog.

The novel is read as a satire of the New Soviet Man, Soviet communism's insistence that the system would transform its male citizens into supermen, men who would no longer fall prey to their natural impulses but instead achieve self-mastery. One of the central characteristics of the New Soviet Man is a selfless devotion to the collective, which might manifest itself in the form of self-sacrifice. The New Soviet Man is willing to give his life for his nation's cause or any cause furthering the glory of his nation. The novel was banned from publication in the Soviet Union but circulated illegally until 1987, when it was finally released. *Heart of a Dog* is now read and discussed widely in Russia and many other countries.

Like Sharik and the New Soviet Man, Laika is an experiment of the Soviet state, and she gives her life to that experiment. While we cannot say that Laika understood what was happening to her or that she chose to be a space dog, and while certainly she did not choose to be sacrificed, like most dogs she was devoted to her caregivers and trainers. What they asked of her, she did. Laika was selfless because she was a dog; it is humans who are selfish and self-serving, who ask animals to accept risk and danger in their stead. Fundamental to Laika, and the other space dogs, is her innocence. "Man, do not exalt yourself above the animals: they are sinless," writes Dostoevsky in *The Brothers Karamazov.*

Much like Philippovich in the novel, Soviet scientists experimented with dogs, killing them and bringing them back to
life. A 1940 film, *Experiments in the Revival of Organisms*, apparently filmed at the Institute of Experimental Physiology and
Therapy in Moscow, features Doctor Sergei Bryukhonenko's
autojektor, a pump not unlike those systems used today in renal
dialysis and extracorporeal membrane oxygenation. The film
begins with a demonstration of the autojektor keeping organs
from a dog alive outside the body, the heart and lungs, for
example. Following that, the scientists reanimate the severed
head of a dog, again using the autojektor to circulate blood.
When a technician stimulates the eyes of the severed head, it
blinks. An application of citric acid to the isolated head's nose
and mouth causes it to lick its lips. The head also reacts to
bright lights and to sound, a hammer banging away at a table's
surface. Next comes the whole dog. Under anesthetic, the scientists drain the blood from a dog's body. The dog dies and
remains dead for ten minutes. They put a stopwatch on it. The
scientists then pump the blood back into the dog and keep the
pump running until the dog resumes normal heart and lung
function. In time, the dog gets up off the table and resumes
its life. The lab where these experiments were conducted, so
the film claims, kept a number of dogs in residence, many of
which had been previously killed and reanimated, with no visible ill effects.

Human beings are animals with a central anxiety that
grows out of our knowledge of the limits of earthly life. Death,
and our fear of death, is a source of deep and perhaps nearly
insolvable loneliness. To be alive is to be at war with loneli-

ness. Experiments like the reanimation of dogs rose out of a near worship of science and technology in the Soviet Union (and elsewhere) and made possible the hope for a future where human beings could mitigate the process of aging, even overcome it. Such experiments, writes Henrietta Mondry in *Political Animals,* "were perceived by the scientific community and the public alike as a step towards the possibility of immortality and resurrection." Laika and the space dogs' "journey to the stratosphere," writes Mondry, "propelled them into the domain of celestial beings. It gave them immortality that could be understood metaphorically but also literally." So too with Soviet cosmonauts, who are the embodiment of the New Soviet Man. After becoming the first human in space, Yuri Gagarin was awarded the title of Hero of the Soviet Union, the nation's highest honor.

◌

Mukha, or Little Fly, served as the control dog for Laika. She might have served as Laika's backup—the role Albina played—but the team felt she did not have the appropriate conformation. Her legs were too short, her body out of proportion. She was not a fitting symbol for the Soviet Union. How could such a homely dog become the darling of the world? So Mukha would stay on the ground, but as the control dog she would train no less vigorously. She would do everything Laika did in preparation for the flight. She would live in the same conditions. She would undergo the same training exercises. She would eat the same food. She was Laika's shadow. The data collected from Laika during her time in orbit would later be compared to data collected from Mukha on the ground. Mukha would also test the life-support systems in Laika's capsule before launch. If she could live sealed inside the capsule on the ground, then Laika could live sealed inside the capsule in space.

When the day of the capsule test arrived, Mukha went into the little slot where Laika soon would go, wearing the suit attached to restraining chains, and facing the window where that stream of light came in. For the next three days all of Mukha's air, food, water, and waste elimination needs would have to be provided by the capsule.

Imagine that little dog sealed inside the capsule. How long before she got bored? Is "bored" even the right word for what a dog experiences in its mind? Agitated, sure. Restless, of course. Her stomach desperate for reasonable food, certainly. Even spiders in space (dining on filet mignon) were fed better than Mukha, and later Laika. A day passed. And another. Mukha seemed to be doing just fine, but she was not eating her food. It was probably hard to tell, however, peering in through the window at that lump of sticky goo in the feeding tray. Had she eaten a little of it? None of it? Half of it? On the third day one of the team members, Konstantin Dmitrievich, made a routine check and found Mukha in distress. He "looked inside the cabin through the window," Ivanovsky writes in *The First Steps*, and "saw such sad dog eyes full of tears that he became uneasy." The team made the call to bring her out. Once out, they discovered that Mukha had not touched any of her food, which meant that she would also have had no water. After nearly three days in the capsule she was severely dehydrated and possibly on a downward spiral toward death.

"It remained unclear how the doggy had lived these days," writes Ivanovsky. Inside the capsule she "did almost nothing but breathe." This was strange behavior for Mukha, who had always performed well during her training. According to Ivanovsky, some of the scientists and engineers made light of it, explaining that "short-legged Mukha was upset to learn that it was not her but long-backed Laika who had won a flight into space." A humorous anecdote, to be sure, but the incident might better serve here as a warning, evidence that once in space, Laika would not fare much better. Ivanovsky notes that Konstantin Dmitrievich had a "rather serious conversation" with a colleague about Mukha's trial run. They approached the problem from several angles: Mukha's personality, a misunderstanding of dog psychology, the space dog food. The food had been prepared by biologists who assured the team that it was adequate. But how did it taste? What Konstantin Dmitrievich knew for certain was that Mukha did not eat, even after three days with nothing else to eat. So "why not just flavor this food with delicious sausage?" Ivanovsky writes.

◻

In some cultures, it is believed that the dog is a medium through which communication with the spirit world is made possible. Anubis in ancient Egypt attends the dead, oversees mummification, and measures the scales that weigh the heart to determine if its bearer is worthy of entering the land of the dead. The dog is a guide between the worlds of life and death, between the known and the unknown, the human and the animal (for the dog reminds us that we are animals), between the conscious mind and the murky wild of the unconscious mind, between the psyche and the soul. The dog is our companion, just as death is our companion. The dog does not desert its master even in death; the two might travel into death together. The dog eats the dead and chews the bones, and digs to exhume or to bury the dead. Anubis is the protector of graves. The dog guards the gates of hell—Cerberus (Greek) and Garm (Norse), the hellhounds, or the hounds of hell. The dog is a healer. The dog is a wanderer too.

The human animal longs for the companionship of the animal dog, for connection with dogs is to reconnect with our animal natures, our animal soul. The human animal longs for companionship with dogs to help us dismember the ego and join in the feast of the senses. This is why the Norse god Odin travels with companion wolves; why dogs accompany the war god Mars; why the founders of Rome—Romulus and Remus—are said to have been raised by wolves; why in North America the wolf is revered for its courage, endurance, cooperation, and protection of its young; why Fenrir, the death wolf of Nordic myth, destroys the old to make way for the new.

The dog is an extension of the human animal, like a club is an extension of the hand to empower it. We cast a dog out beyond us to test the boundaries of our sight, to take us where we cannot go alone. We cast a dog out beyond us to scout. Dogs are scouts. Hands and eyes, the shepherd sends his dogs out to bring the sheep around, or bring the lambs home, to defend the sheep and lambs from the wolf, the bear, the big cats. On the hunt, we send dogs out to retrieve what we have killed. We send dogs out to kill. We send dogs out to tell us what is out there that we might kill and eat, and so survive. The dog improves us, enhances us, makes us stronger. And it has long been so. The dog's unparalleled sense of smell, with its ability to track and discover, can help us find what we have lost. We cast a dog out on the trail of our own longings, our searches, our explorations. The dog does not hesitate, or waver, or falter. It is singular in its purpose, its only life stripped down to what is, at last, necessary. The dog is an extension of the human mind, like science is an extension of the human mind. We cast the dog out before us like a scientific instrument, like the Mars rover *Curiosity*, an extension of our hands and eyes on a far distant planet. The Mars rover is like a mechanical dog. Dogs can understand us, our feelings, our gestures, our language. They can nearly speak to us, at least we can understand something of what they want to say. And in listening to what they say to us, and what we say to them, we come to think of them as feeling like us. We sometimes think we feel like them. And a dog will go to its death for us, to protect us, to protect its place that is with us.

What is all of this but an expression of love? Dogs will go and do and give, to us, because we ask, because they depend on us, because they love us, because we, in turn, love them.

FOUR

◌

Scouting the Atmosphere

Let the dog, man's helper and friend since prehistoric times, be
sacrificed for science. But our dignity obligates us to do this only
when necessary and always without unnecessary torment.

IVAN PAVLOV
inscribed on the memorial to his lab dogs in
Saint Petersburg, early twentieth century

Between 1951 and 1966, the Soviet Union launched some for-
ty-two rockets into space and into orbit carrying more than
fifty different space dogs. Most flew in pairs, and with the exception of two dogs that each flew with a mannequin, Laika
is the only dog who flew alone. Some dogs flew only once,
because they were either killed or retired for some reason.
Others flew multiple times, the record probably held by a dog
named Otvazhnaya, or Brave One, who flew seven times.

Cataloging the flights of the space dogs is a difficult task,
because some had more than one name. In fact, some of the
scientists and engineers working on rockets and missiles also
had alternative names to help hide their identities. If a dog
died on a flight mission, a new dog might be given its name.
Some of these dogs assuming the name of another were numbered in sequence, as in Fox and Fox 2, and some were not.

The team might change a dog's name from its first to its second flight, or change a dog's name as it waited on the launchpad for liftoff. The two most reliable sources I found for the identification of the space dogs and their missions are Olesya Turkina's *Soviet Space Dogs*, which includes a comprehensive list of space dog flights; and *Animals in Space,* by Colin Burgess and Chris Dubbs, who invested considerable time and effort in piecing together a chronology of the flights.

To better understand Laika's story, a number of space dog flights deserve our attention, missions by some dogs who flew before her and some who flew after.

◻

Three hundred fifty kilometers up the Volga River from its delta at the Caspian Sea is the site of the Soviet Union's once secret military missile test center, Kapustin Yar. Backed by the vegetated green of the great Volga's braided channels, the test center faces eastward onto a vast system of dunes and wastes, the Ryn Desert stretching some six hundred kilometers into western Kazakhstan. Temperatures here can reach highs of 45 degrees Celsius in summer and lows of −35 degrees Celsius in winter. It is a windswept and arid land, empty of people but for a few scattered villages and towns. Kapustin Yar is not so secret anymore, but at its genesis in 1946 it was central to the Soviet Union's covert missile program and the site of the first suborbital space dog launches.

On July 22, 1951, the first two space dogs—Dezik and Tysgan—were launched into the upper atmosphere from Kapustin Yar. Laika had yet to be born, and while putting a satellite into orbit, followed by a dog and eventually a man, was the dream of some engineers and scientists, it was yet far from possible. So much was unknown. So much had yet to be tested.

Mostly white with medium to long hair, Dezik had a fluffy, kind-looking face and an amiable and positive spirit. Tsygan, whose name means "gypsy," had a mostly white body with black patches over her eyes and much of her head, like a border collie. Both dogs had remained calm and easy during the rigorous training, making them prime candidates for this first test flight.

While the rocket that was to carry Dezik and Tysgan into space waited on the launchpad, the dogs were dressed in their flight suits and sealed in a capsule inside the rocket, even the capsule that would be repurposed for Laika's flight six years later. The team had chosen to launch early in the morning because the light at this hour would make the rocket visible against the broad expanse of desert sky. A fleet of vehicles sat nearby so that the recovery team could drive to the site where the capsule came down. Also invited to the launch was a party of dignitaries, VIPs, and members of the Soviet government, all eager to witness this historic launch. Just before the hatch was closed, Yazdovsky spoke to the two dogs through the portal, words commonly offered to Russian soldiers on their way into battle. "Return with victory," he said.

The rocket went up, a long line of white into the blue. A million-horsepower engine pushed the rocket higher and higher, hitting a top velocity of 2,610 miles per hour. The dogs' heart rates shot up. The g-force of acceleration bore down on them, their bodies now weighing five times more than on Earth.

They could not hold up their heads. The rocket reached an apex altitude of more than 100 kilometers or 62 miles, just above the Karman line. And then a sudden and calming relief as the dogs entered microgravity. Their heart rates slowed and returned to normal. Secured inside the capsule, Dezik and Tysgan did not float about the pressurized cabin, but they were nearly weightless for four full minutes.

On the ground, the party of onlookers watched, as in the distance an object could be seen falling back to Earth. It struck the ground hard and exploded in a ball of flame. It was not the capsule, however, but the cast-off rocket body, the remaining fuel igniting and burning on impact. The spacecraft that carried the dogs was still in the sky, turning over the smooth curve of its path, and then down it came, faster and faster. A white parachute appeared in the sky. At this distance, the observers watched the chute like a tiny paper toy unfurling in the blue, slowing the descent of the spacecraft, slowing it, as down, down it fell.

Yazdovsky had directed the party to remain at a safe distance until the spacecraft landed, but the weirdness of it, the novelty, the joyful excitement was too much to bear. Two little dogs were inside that spacecraft, and they had just made a trip to outer space. Were they alive? Were they injured? Was their air running out? What was it like to ride that rocket into space? As the team ran for their cars, the invited guests ran after them, and in a torrent of gas and dust the whole lot went shooting across the desert to the landing site.

Alexander Seryapin approached the spacecraft first. He was especially fond of Tysgan, he admitted in an interview for the film *Space Dogs*, and now that she had been to space and back, he was desperate to know what had become of her. He leaned in to look through the window into the capsule. "Our animals were alive," he said in the interview. "They were sitting calmly. We released the dogs and took off their sensors. We gave them sausage to eat, and water. Everyone was very happy, especially Sergei Korolev. He was normally such a serious man, so I was surprised when he grabbed one of the dogs and ran around the [spacecraft]." All the reports concur that Dezik had no identifiable injuries or ill effects from her ride, and Tysgan had but a scrape on her belly where the capsule had caved in against her upon landing. While the US was just completing the Albert flights and killing one monkey after another, the Soviet Union had launched its first dogs into space and recovered both alive.

A week later Dezik flew again into space, this time with a dog named Lisa, Russian for "fox." The rocket went up, the spacecraft came down, but this time no white chute appeared above the horizon. The spacecraft kept falling until the Earth came up hard beneath it. Reports say that a pressure sensor damaged by the vibration of the rocket engines prevented the braking chute from deploying, a mechanical failure that killed both dogs.

When Anatoli Blagonravov (1895–1975), the head of the Commission for the Investigation of the Upper Atmosphere, heard about the death of Dezik and Lisa, he grieved deeply. Blagonravov would later become a towering figure in the Soviet space program. Among other achievements, he was instrumental in negotiating the 1972 agreement between the

US and the Soviet Union that called for the development of spacecraft built by the two nations that would be capable of docking with each other. The 1975 Apollo/Soyuz Test Project (known as Soyuz/Apollo in Russia) is often cited as the end of the Space Race and so the beginning of cooperation in space between the US and USSR. But that hallmark event was still down the road. Blagonravov was crushed by the death of Lisa and Dezik, especially Dezik. He could not stand to think of Tsygan, Dezik's former flight mate, flying again and maybe dying in a crash too. He announced to the team that Tsygan would retire and then promptly adopted her. According to Yazdovsky, Tysgan lived a long and easy life in Blagonravov's Moscow home.

You might imagine them—this man so central to the Soviet space program and this little dog, one of the first in space—taking long lovely walks through the Moscow streets, maybe into Gorky Park, passing by the public fountains there, and the many people at their leisure.

○

Ryzhik (Ginger) and Smelaya (Courageous) were scheduled to fly out of Kapustin Yar on August 19, 1951. The day before launch, Smelaya cut loose from her handlers during her walk and ran off into the barren steppe. She bore similar markings as Tysgan (now living happily in retirement in Moscow), white with black patches over her eyes and much of her face. Her ears stood up, giving her the distinct look of a dog close to her wild ancestors. Out there in the steppe where Smelaya was last seen running, her wild ancestors waited for her, the golden jackal, an opportunist that would happily prey on a wayward space dog. Still, it was not unheard of for dogs to carouse with jackals, as hybrid jackal-dogs too roamed the desert. Summer temperatures might be hot during the day, but it would cool off at night. Still, the team came to understand that it was unlikely Smelaya would return. It was bad enough to lose her to the desert, but what complicated matters further was that space dogs were selected in compatible pairs for their flight missions. How would they find a replacement on short notice?

The next morning, launch day, Smelaya walked back into Kapustin Yar as if it were the most ordinary thing in the world. Where had she been? What had she been doing? How did she make it through the dangerous night? Whatever the answers, the team prepped and dressed her for flight, she flew, and she was recovered safely. Smelaya is likely the only dog in history to spend a night with jackals and fly into space the next morning.

◻

Yazdovsky chose a dog named Neputevyy (Screw Up) and another named Bobik to fly on September 3, 1951. Like Smelaya, Bobik escaped from her handlers on a preflight walk and vanished into the steppe. The previous return of Smelaya gave the team some hope, but Bobik did not return and they were out of time. As the flight schedule demanded a replacement dog, Yazdovsky ordered that one be found. A lab technician suggested they fly one of the many stray dogs hanging around the mess hall at Kapustin Yar. The team agreed that this was not a bad idea. "So I put on a raincoat and off we went," Seryapin said in the film *Space Dogs*. "We caught one of the dogs, about the same size as the runaway. We brought him in, washed, fed, and brushed him. [It isn't certain whether the dog was male or if Seryapin is defaulting to the masculine pronoun.] We tried the sensors on him. The dog was absolutely calm. While we worked on him, he licked our hands. He was a very calm dog." Even so, this calm dog had no training at all, and the team worried the flight alone might mean his death.

The rocket went up and came down just fine, and what about the dogs? "I looked into the hatch," Seryapin said. "You know, I think my heart missed a beat. Both dogs were alive, and the new dog was absolutely calm.... When [Korolev] saw I was holding the new dog, he asked, 'But which dog is this? Where did it come from? Why have I not seen it before? What's its name?'" Seryapin explained what had happened. "Korolev petted the dogs as he always did," Seryapin said, "and then he said, 'Remember comrades that a time will come when our trade unions will offer ordinary people holidays in space. Well, here's the first one.'" And he held the dog up in triumph.

Korolev named that dog ZIB (a Russian acronym for *Zamena Ischeznuvshevo Bobik*, or "the replacement for disappeared Bobik"). Blagonravov, perhaps much impressed by ZIB, took him home too, where he lived in happy retirement with Tysgan.

✺

On June 26, 1954, Lisa 2 (Fox) and Ryzhik 2 (Ginger) flew into space. Both dogs were recovered alive, but this flight, and the next eight flights, included some major changes. First, the program had been making improvements to the rockets, which were now more reliable and more powerful. Second, the recovery system, one of the weaknesses of the first series of dog flights, had changed. The braking chutes sometimes failed, resulting in two dead dogs. Even when the braking chutes did work and the dogs were recovered, the team sometimes found blood spattered on the inside of the capsule walls and coming from the noses and anuses of the dogs. A dog named Damka flew twice in the same week, and the team found hemorrhaging from her eyes. These dogs recovered, but the team came to understand that the g-force associated with braking was taking a toll on the dogs.

In this next series of flights, instead of flying in a sealed capsule, the dogs would wear a pressurized space suit with a helmet and ride in a protective metal framework, or chassis, that would be ejected from the spacecraft at different altitudes during the descent to the ground. The flights would go like this: the rocket would ascend to an apex altitude of sixty-two miles, the very edge of space. Here the nosecone of the rocket would separate from its booster. On its descent, the first dog would be ejected at fifty miles altitude, and the chute would open at about forty-seven miles altitude, giving it a long, mostly gentle ride back to Earth. The nosecone would continue to fall, and at twenty-eight miles altitude the second dog would be ejected. Its chute would not open until about two miles altitude, leaving the dog to freefall for a full twenty-six miles, straight down. The chute on the nosecone itself would open before that, at about four miles altitude.

For the first time, telemetry would be used to relay information to the ground crew, especially the space dogs' heart rate, blood pressure, respiration, and urine output, if any, as well as the temperature inside the space suit. The oxygen supply was to last two hours, and at 12,000 feet the dog would move off the bottled oxygen when a valve opened to allow breathable air to flow into the helmet.

○

Despite steady improvements, rocket flight was dangerous, and a number of space dogs died. Lisa 2 (Fox) flew again on July 26, 1954, before she was killed in a flight on February 5, 1955. Her partner on the flight, Bulba, was also killed. Fox was an all-white dog with longer, soft-looking fur and a friendly face. File photos show her right ear standing tall while her left ear is cocked to the side. She was a favorite of Alexander Seryapin, and reciprocally he was a favorite of hers. "When [Fox] went about with [her] guard it was best to keep your distance," Seryapin said in *Space Dogs*, "otherwise the guard would snap at your trouser legs." The team around Kapustin Yar laughed about that, such a little dog with a big bite, always guarding Seryapin. Annoying sometimes, sure, but endearing too. Then "a time came when I had to put Fox into her space suit and send her on a flight," Seryapin said. "I was well aware an animal shooting up to a height of 110 kilometers doesn't bear the stress so well. So I put Fox into the [spacecraft] so she could see the Earth while descending from 90 kilometers. Unfortunately the rocket veered to one side, and when the rocket tried to correct itself there was a sharp jolt that threw Fox out of the [spacecraft] into the atmosphere in her space suit. The jolt was so strong that the animal was dead before she fell to Earth."

The program's secrecy forbade memorials for the dogs, but Seryapin defied this regulation to bury Fox in the wild steppe. He photographed her gravesite. Did he intend to return one day to honor her, to sit beside the little mound of her tucked up against a hummock in the shade of a saxaul thicket? Did he intend to steal away from his barracks in the night to visit her, his silhouette on a hilltop dune going over it, and off into the empty lands, the bark of golden jackals coming up for him out of the distance? Did he have words to say when he laid her in?

◻

Spaceflight and exploration were the greatest challenges in aviation medicine, and naturally the field attracted some of the best scientific minds. In summer 1956 medical doctor Oleg Gazenko joined Yazdovsky's space dog team at the Institute of Aviation Medicine. A thin man with a neat mustache and sharpened nose, Gazenko had worked for the Soviet air force during World War II. After the war, he began to work on the issues that affect pilots at high altitude. The military had just constructed a larger, more powerful rocket, the R-2, which could boost heavier payloads to higher altitudes. Instead of just sending dogs to the edge of space (62 miles), the R-2 would take them to an altitude of at least 130 miles. The dogs had to be better trained, and Gazenko was one of the men who would train them.

To learn as much as he could, Gazenko made a trip to the circus to speak with animal trainers, likely the Old Moscow Circus on Tsvetnoy Boulevard. As one of the oldest circuses in Russia, trainers at the Old Moscow Circus would have been well acquainted with the methods of the famed Durov family. Vladimir Durov (1863–1934) was long dead, but his training methodology and legacy lived on in his family and others who worked with circus animals. Durov was widely known in the Soviet Union as the man who revolutionized animal training by using kindness and love instead of force and punishment. His "elevated status is based on his reputation as a friend to all animals and as an operator who used kindness to make them do his bidding," writes Henrietta Mondry in *Political Animals: Representing Dogs in Modern Russian Culture.* "Indeed, Durov is the father of a humane method of animal training. His brand of education using love and patience...has produced extraordinary results." Of all animals, Durov loved dogs best and regarded "them as his friends, and the animal closest to human beings," writes Mondry. "For Durov, the correlation between humans and dogs was an established fact."

Korolev and Yazdovsky had embraced a humane approach to training space dogs from the beginning, and now Gazenko would refine it. One of the initial concerns about using dogs to test life-support systems for spacecraft was their individual natures, that no two dogs were exactly alike. This meant, for example, that one dog's reaction to g-force loads would not necessarily be the same as another. For this reason the team flew two dogs at a time, so that their responses might be compared. In designing their training, Gazenko assessed each dog's personality and trained her accordingly. "We were more interested in preflight training than in biological experiments," Gazenko said in an interview with the Smithsonian in 1989. "Instead of concentrating on the body, we were more interested in the creature itself, the dog's personality. So we observed their behavior and perhaps learned the principles we used later in the selection and training of cosmonauts." In all my reading, I found little attention to this key detail. Not only did the space dogs test the hardware required to make human space travel possible, but they also helped develop the training regimens, along with an understanding of the personality traits crucial to becoming a cosmonaut. Dogs and humans achieved spaceflight together.

Along with the animal training methods of the Durov family, Pavlov's development of respondent conditioning is deeply seated in the way the Soviets worked with animals, not as a policy so much but as a culture. Instead of punishing the space dogs to get them to perform or not perform, the Soviets created an environment where the dogs performed willingly. Respondent conditioning, which Pavlov famously refined using dogs, pairs what is known as a potent stimulus (food, for example) with a neutral stimulus (a bell). When the bell rings, the dog gets food. Since food generally elicits salivation in dogs, and the dog comes to associate the bell with food, soon the dog will salivate upon hearing the bell. No food is required.

Alongside Pavlov's respondent conditioning is American psychologist B. F. Skinner's (1904–1990) operant conditioning; taken together they give us the principles of behaviorism. Operant conditioning is divided into two branches: one for reinforcing a behavior, which is to increase it, the other for punishing a behavior, which is to decrease it. Both branches include options for positive and negative stimuli.

To increase Ham's success at the control panel, American

trainers used negative stimuli (electric shock). To avoid this shock, the chimps learned to operate the levers correctly. Such a training method can be effective, and it was in the case of the space chimps. While Dittmer tells us that the chimps were well cared for in their living quarters and during training, it is interesting to consider the reward-based training of the Soviets next to the punishment-based training of the Americans. What does this difference mean next to the American belief that the Soviets were cruel and godless, a nation built on the foundation of Stalin's terrifying reign? Is it appropriate to regard the USSR as a nation of people controlled by punishment, while its research animals performed for reward; and the US as a country where those roles are reversed? If so, what does this say about human empathy? Which path is an expression of empathy, if either is at all? At the risk of sentimentality, if the Soviets used love to train their space dogs, then a trained space dog, like Laika, completed her daily work not out of the threat of being punished but out of love.

◻

In 1955 the Soviets began construction of a new spaceport in Kazakhstan about five hundred miles east of Kapustin Yar, at the edge of the Kyzl Kum desert. This new, modern facility—now known as Baikonur Cosmodrome—was the largest spaceport in the world, and it remains so to this day. All crewed missions launched from present-day Russia fly out of Baikonur, including those carrying American astronauts. It was here that the team tested and perfected the R-7 rocket, the world's first ICBM, the rocket that carried *Sputnik I*, *Sputnik II* and

Laika, and the first human, Yuri Gagarin, into orbit. Baikonur was built in part to accommodate the R-7 rocket. It was just too big and powerful to be launched from Kapustin Yar. And so successful was the R-7 that the Soviets used its basic design as the foundation for a whole family of rockets. To date, more R-7 rockets have been launched than any other comparable family of rockets in the world. The spaceport is known colloquially as Gagarin's Start, but it is also sometimes called Tyuratam because it is close to the village of Tyuratam. In time, a town grew up out of the desert to serve the spaceport, and it was named Leninsk. In 1995 President Boris Yeltsin changed the town's name from Leninsk to Baikonur.

Like Kapustin Yar, Baikonur was a desolate place—cold, wind-battered, and isolated. It was about as far from the glamor and beauty of Moscow's Bolshoi Theatre or Leningrad's Church of the Savior on Spilled Blood as a man could get. Yet it was also a place of great activity, energy, and excitement, as some of the world's most secret and most advanced technologies were being developed and tested here. Not long after the spaceport's completion, the R-7 rocket carrying Laika shattered the morning quiet, marking a major turning point in Soviet, and world, space exploration. From that moment forward, rockets and their spacecraft would no longer be flown solely to the edge of space but into Earth orbit and, following that, to the moon, and then deeper and deeper into our solar system.

□

Some three years after Laika's flight, Lisichika (Little Fox) and Chaika (Seagull) were scheduled for an orbital test flight on an R-7 rocket in the prototype of the new Vostok space-craft. The flight, scheduled for July 28, 1960, was a critical test of the Vostok, which in about a year's time would take Yuri Gagarin into orbit. Khrushchev had explained to Korolev that he would be held personally responsible if an American astronaut was first into space. You can imagine what was riding on these two little dogs: Korolev's reputation and place in the pantheon of Soviet heroes, the space dog and the rocket program itself, and the world's regard for the power and position of the Soviet Union. In effect, everything.

Korolev, the man whose keen focus and hard edges were legendary, whom most of his engineers and scientists feared, kept a quiet fondness for Lisichika. She was one of his favorites. How this unfolded we can only imagine. Perhaps it arose from a singular incident when Lisichika was first brought into the kennels and needed someone. Perhaps Korolev gave her food and water on one of his morning rounds, and she gave him a wagging dance and a soft look about her eyes, which softened him. Or perhaps over time she warmed to him, and he to her. We cannot know. Despite his fondness for Lisichika, Korolev regarded her as a highly trained space dog, a working animal. She was not a pet. In fact, his love for her may have stemmed, in part, from what she could do for him, what she could do for the Soviet Union. The investment in her had to be returned to the program and to the country, and so she would have to fly. On the day that Lisichika was scheduled to fly, Korolev bent to her ear and whispered: "It is my deepest desire that you come back safely."

Waiting on the launchpad inside the capsule, Lisichika and Chaika felt the vibration and storm of the engines igniting, their burning roar, readying to lift that rocket into space. In that critical moment, as the team waited for the rocket and the dogs in which their lives were so invested to rise into the heavens, a strap-on booster broke loose and fell away. The rocket exploded on the launchpad. Both dogs were killed.

▫

Inside the Memorial Museum of Cosmonautics in Moscow are two glass cases housing the dead and stuffed space dogs Belka and Strelka. The cases are positioned on either side of the capsule inside which they rode into space. The capsule is dented near the bottom where it impacted the ground. Belka sits on the right side of the capsule with her pointy nose, her mouth in an eerie grin, her toenails too long. Her head is positioned upward and at an angle, so that upon approaching her, she seems to look beyond you, into a distant corner of the ceiling. Strelka, on the left side, is seated too, almost crouched. Her eyes gaze upward as if looking at the person standing over her or perhaps looking at the stars. In *Soviet Space Dogs*, Turkina writes that likenesses of Belka and Strelka are often positioned this way, "looking toward the heavens" because their "portrait evokes the iconography of the heroic human pilots portrayed in Soviet paintings and posters: always moving upward, always toward the sky."

During my visit to the museum, I stood for some time before the capsule and the dogs. I had read about them, read the details of their story, and I had looked closely at photographs of the dogs taken before and after their flight. While they were but objects now, preparations of remnant skins, I could sense something of what the dogs had been. They came to life for me in a way that fails a photograph. Standing in their presence, they were no longer lost in a past I had not lived but were part of a present I was now living. Somehow, standing there, I felt closer to Laika too, who had gone out of this world and come back into it, both times in a blaze. Like me, Laika had been separated from Belka and Strelka by time, even if only a few years. Yet the way time pools in such a moment in the presence of the dead, you can believe without proof that the threshold of time, both into the past and into the future, is possible to reach across.

After Laika, Belka and Strelka are the most famous space dogs of the Soviet program. They achieved what Lisichika and Chaika did not: they became the first living beings to orbit the Earth and be recovered safely. They proved that the life-support systems on the Vostok and the newly developed hardware for deorbiting a spacecraft worked, and that the USSR now possessed the technology and know-how to put the first human being into space.

Belka (Squirrel) weighed just twelve pounds and measured eighteen inches long and twelve inches tall. She was nearly all white but for her sides and short, stiff ears, which were tinted yellow. Belka's broad, stout-looking head distinguished her from Strelka (Little Arrow), who was more gracile, leaner and longer, though she too weighed twelve pounds. Strelka's head, back, and sides were marked by dark brown patches, and her ears look softer, more pliable, which gave her an amiable countenance. In some photographs, Strelka's ears are laid back or bent down but not floppy like a yellow lab's. In publicity photos and on TV, Belka typically wore a red flight suit while Strelka wore green. Belka had originally been called Albina (though she is not the same Albina who was Laika's second for *Sputnik II*), and Strelka had been named Marquise. Turkina reports that it was the commander-in-chief of the Strategic Missile Force, Marshal Mitrofan Nedelin, who thought the names too French, too bourgeois, and so directed the team to give the dogs strong Russian names.

With the success of their flight, the world fell in love with Belka and Strelka. Their images were widely published. They made radio and TV appearances and were trotted out for display before crowds of citizens and dignitaries. They were featured in newspaper and magazine stories, and their tale of heroism was retold in children's books and more recently in animated films. Like Laika, their images were fashioned into porcelain figurines and appeared on stamps, ornamental wooden boxes, confectionery tins and chocolate boxes, badges, postcards, cartoons, you name it. Fan mail poured in. There was no pop culture in the Soviet Union in those days. "Under socialism the niche occupied by popular culture in capitalist society was subject to strict ideological control," Turkina writes. "Paradoxically, Belka and Strelka became the first Soviet pop stars."

In 1961 Van Cliburn came again to Moscow to perform, for he so loved the city and the nation. While he was recording a live performance at the Shabolovka broadcasting center, Belka and Strelka were featured TV guests in a nearby sound studio. Wily little things that they were, the two dogs escaped their handlers and scurried off through the hallways of the building. "We were at the Shabolovka studios where they were filming the dogs," said Institute of Aviation Medicine biologist Ludmilla Radkevich in an interview in *Space Dogs*. "The cameraman said: 'Could they bark a bit? Make them bark. Do something for the cameras.' But the dogs just sat quietly until one of the cameramen dropped something. Then the dogs jumped up and ran right out of the studio." The two dogs slipped through a doorway and found themselves on stage with Van Cliburn. The great pianist recognized the dogs immediately and interrupted his concert to welcome them. "He was so delighted and happy he couldn't believe his luck," said Radkevich.

Van Cliburn posed for photographs holding the dogs. In one photo, he holds the dogs close against his chest, Belka in his right hand, Strelka in his left, a happy smile on his face, the studio lights putting a shine on his always-perfect hair. In *Soviet Space Dogs*, Turkina states that the incident made the evening news, the American prodigy poised in awe and admiration of the Soviet space dog heroes. The moment is a giant among the many instances of what the Soviet PR machine called "Victories of Soviet Science for the Sake of the Entire Human Race."

Belka, who had already flown three times, and Strelka, who had never flown, made their historic flight on August 19, 1960. The team prepped the two dogs, attaching sensors to monitor heart rate and respiration. Their orbital flight was scheduled to last twenty-four hours, so their space suits included a waste management system, and their capsule, a food dispenser. Sensors allowed the ground crew to monitor the air quality inside the capsule, its carbon dioxide, oxygen, and water vapor levels, while TV cameras monitored the dogs in real time. Also along for the ride: rats, mice, insects, fungi, various plants and sprouts of wheat, peas, onions, and corn, and, according to some sources, a rabbit.

The team watched the video stream of the dogs in the capsule as the R-7 rocket lifted off. Belka and Strelka were pinned down by the g-force of acceleration. They looked almost dead, or dying, it was hard to tell. Their heart rate and respiration rose dramatically, even tripling, which let the team know the dogs were still alive but obviously in distress. And then the spacecraft entered orbit and that soft elevation of microgravity collapsed over them. Their vital signs normalized. They began to stir, to move inside the capsule a little, and then Belka started barking. The team on the ground could not hear her, as the microphones they installed were only able to detect the background noise inside the spacecraft, but they could see her. She struggled and barked in that weird absence of gravity, nothing like the environment her body had evolved in. So she barked. A dog barking in space. The spacecraft came around the Earth for the fourth time, and Belka vomited. That done, she settled down and seemed to accept what was happening to her. As her body adjusted to microgravity, she went along with the ride. There was nothing else she could do.

It's likely that Belka's distress and vomiting were caused by what researchers now call space adaptation syndrome, the effect of moving rapidly from the hypergravity environment of launch to the microgravity environment of space. This shift causes an increase in cranial pressure, which can lead to headache, vertigo, a general feeling of malaise, and vomiting. The condition affects about 50 percent of all astronauts who fly in space. Yazdovsky and his team were observing these effects for the first time and acknowledged that there were just too many unknowns about microgravity. They recommended the first human in orbit make no more than one trip around. The recommendation held, and this is precisely what Gagarin did.

After eighteen orbits in a period of some twenty-five hours, the spacecraft carrying Belka and Strelka made its reentry into Earth's atmosphere. The capsule the dogs rode in was ejected from the spacecraft and came down under a braking chute. It landed off course in the steppe near the Russian city of Orsk at the southern tip of the Ural Mountains, that great sweep of temperate grassland where, some six thousand years ago, the horse was first domesticated, revolutionizing the way we live and work, hunt and fight. The recovery team found the capsule and opened it. Inside were Belka and Strelka in perfect condition. Once released from their restraining harnesses, the two dogs ran about the search and recovery team barking and leaping into the air.

Belka and Strelka were interviewed on Radio Moscow and put on display at a press conference at the Academy of Sciences. In a now-famous photograph, Gazenko holds the two dogs aloft, one in each hand, the look of triumph in his smile and eyes, a lighted candelabra in the foreground. There was no doubt, in that moment, that Belka and Strelka's success would not have been possible without Laika's death.

In November of that year, Strelka gave birth to a litter of six puppies sired by Pushok, a space dog who never flew. Khrushchev gave one of the pups, Pushinka (Fluffy), as a gift to America's first family, the Kennedys. Pushinka was a gift of political and cultural goodwill, but also a kind of gloat: see what the Soviet Union has achieved! Before joining the Kennedy family in the White House, Pushinka endured a thorough inspection, which included X-rays, to make certain she was not bugged or surgically implanted with some explosive device. Little Pushinka, the Russian debutante, later fell in love with one of the Kennedys' dogs, Charlie. They had a litter of pups, American-Soviet mutts the Kennedys named Butterfly, Streaker, White-tips, and Blackie. President Kennedy called them pupniks.

◻

With the flight of Belka and Strelka, the space dog program was coming to a close. The Soviets knew it would not be long before they sent a man into space, and then the kind of work the dogs were doing to test life support and recovery systems would not be needed anymore. This was not, of course, the end of biological and medical research in space using animals. Such research is ongoing, but cosmonauts would soon become both scientist and test subject. Another factor pushing the end of the space dog program was a growing criticism by both animal rights organizations and the general public. People just didn't want to hear news of flying dogs, and especially news of those dogs dying. Fruit flies? Mice? Fungi? No problem. But dogs, the animal we love best, our companions and friends? Too cruel. Not ethical. Inhuman.

Mukha (Little Fly) and Pchelka (Little Bee), along with

guinea pigs, rats, mice, fruit flies, and some plants, lifted off on December 1, 1960, aboard *Sputnik IV* for a flight that was to last one day in orbit. Mukha had already had two near misses on her life: she had been passed over for Laika's doomed mission, and she nearly died during her three-day test of Laika's capsule. Now she was going to risk it all again. During the flight, TV cameras onboard the spacecraft broadcast a signal back to Mission Control at eighty-three megacycles. The CIA demodulated the signal and watched footage of the two dogs during their flight. It was becoming harder and harder for both the US and the USSR to keep their programs, and their technology, secret from each other. When the spacecraft reentered Earth's atmosphere, the retro-rocket, designed to slow and guide its trajectory, malfunctioned. It fired and kept firing, and it would not shut down, sending the spacecraft far off course. Mission Control was no longer in control, and it looked like Mukha and Pchelka would land outside the Soviet Union. To avoid a foreign government getting ahold of their spacecraft and their dogs, Mission Control destroyed it with a remote self-destruct feature, sending a fiery trail across the sky. Both dogs were killed.

❑

December 22, 1960. The pilot of an Anton-2 aircraft sent a radio message to Arvid Pallo, head of the space dog search and recovery team: "I can see a sphere with two openings. There is also a parachute," he said.

Pallo and a colleague boarded a helicopter and flew to the crash site, about sixty kilometers west of a town called Tura, in far Siberia. It was late in the day, and at this latitude in winter, daylight lasted only six hours. Pallo opted to risk getting caught in the dark and cold because if he couldn't find the spacecraft and disarm its self-destruct system, it was going to blow up, taking the two dogs—Shutka (Joke) and Kometa (Comet)—with it. When the helicopter landed, Pallo and his colleague jumped out, sinking into waist-deep snow. Unsure which direction to go, they soon lost their way. The big Anton-2 cruised by again, the pilot warning that everyone needed to get back to the base. It was getting dark, and the temperature was dropping. The helicopter pilot would have trouble navigating in this unfamiliar landscape at night.

"I used my radio to [ask]...the An-2 pilot to show us the way by flying along a straight line from the helicopter to the spacecraft," Pallo said in an interview for *Roads to Space*. The pilot did so, and Pallo and his colleague set off in that direction. Arriving at the downed spacecraft, dark and cold coming in against them, Pallo and his colleague had to work fast.

The launch of Shutka and Kometa had gone well, but at the edge of space the third stage of the rocket malfunctioned and failed to push the spacecraft into orbit. Emergency systems kicked in, and the spacecraft separated from the rocket at 133 miles altitude. The team assumed the spacecraft and the two dogs, along with some mice, insects, and plants, would be destroyed when the self-destruct system detected its anomalous trajectory, but it didn't. On the descent, the capsule carrying the two dogs should have been jettisoned from the main spacecraft and come down under its own braking chute. That didn't happen either. On the way down, the dogs hit 20g, a crushing force that could kill them, and landed near the Pod-kamennaya Tunguska River, where in 1908 an asteroid or a comet struck the Earth, destroying two thousand square kilometers of taiga forest. The so-called Tunguska Event remains the largest impact event on Earth in recorded history and has been the subject of many UFO conspiracies.

If the dogs had survived the descent and landing, they were not yet safe. The self-destruct system that failed to detonate during flight had a backup timer: the spacecraft would blow up in sixty hours unless the system was disarmed. There it sat in the snow and ice of the Siberian winter, the dogs inside, the timer ticking.

In *Roads to Space*, Pallo said he volunteered to approach the spacecraft first and try to disarm it. "Go and stand behind a tree while I go up to the spacecraft and disable the self-de-struct system," Pallo told his colleague.

The man refused. "This is my system," he said.

"I'll go," Pallo said, "because I'm the leader of the group."

Again the man refused.

The timer ticking, the two men used matches to draw lots. Pallo lost.

There in the taiga forest, darkness coming on, the temperature at −40 degrees Celsius, Pallo watched from behind the protection of a tree as his colleague approached the spacecraft. The man set to work disarming the self-destruct system, calling out each of the steps to Pallo. The work came along easily, efficiently, and then it was done. But as the dogs' capsule had not ejected from the spacecraft, that system was still armed, as were some of the pyrotechnics from the parachute deployment system. Either could explode and kill the dogs as well as the two men. This time Pallo would do the work while his colleague stood behind a tree. Pallo reached inside the rocket's interior to disarm the system, but his heavy winter gear made it impossible for him to get his arm far enough inside. He removed his coat. As he strained to reach the connector, the spacecraft shifted. "It was anyone's guess what might happen next," Pallo said.

What happened next was that Pallo reached the connector and pulled it free. Now to free the dogs.

"We tried to see the dogs through the portholes," Pallo said, "but the glass was covered with hoarfrost which had built up during the days and nights since the spacecraft was first spotted. We knocked on the walls of the container but heard no signs of life inside." Night was coming on. The helicopter sat waiting, its rotors still turning, burning fuel to keep the engine warm, the pilot beckoning to the men. *Come on. We do not have much time. We must leave now.* Pallo and his colleague left the dogs, Shutka and Kometa, inside the sealed capsule and returned to Tura. If they were alive at all, they would have to endure another long cold night.

Arriving back at Tura, Korolev called Pallo on his radio frequency phone twice, asking for an update on the condition of the dogs and the spacecraft. "As I began to describe our ordeal," Pallo said, "the aurora borealis appeared and cut off our radio communication." The aurora borealis, or northern lights, have been known to disrupt radio and telegraph communication, but they sometimes act as a kind of power source for such equipment. In the so-called Great Geomagnetic

Storm of 1859, a mass solar ejection supercharged the telegraph lines between Boston and Portland, Maine. Operators on either end were able to continue their transmissions with their power systems switched off. But that was not the case on this night. Pallo was cut off, and he had not yet told Korolev that he didn't know if the dogs were alive.

The next morning Pallo returned to the spacecraft accompanied by a veterinarian. If the dogs were alive, they would likely be in poor condition after three nights in such cold temperatures. Perhaps the veterinarian could help save them. As the two men removed the capsule from the spacecraft, they could hear the dogs barking. Pallo opened the hatch to find Shutka and Kometa looking up at him. The veterinarian removed his sheepskin coat, wrapped the dogs inside, and carried them to the waiting helicopter.

Later, Oleg Gazenko adopted Kometa, and she lived out her life in his family home.

○

A little later in China, in 1966, a young male dog named Xiao Bao (Little Leopard) flew into space, followed by a female, Shan Shan (Coral), a couple weeks later. Both landed safely and were returned to the care of their handlers. China had been working with Russian scientists to get its rocket program off the ground so they could conduct biological and medical research. Xiao Bao and Shan Shan were part of that research, along with mice, rats, and fruit flies. But according to Burgess and Dubbs, there is some evidence that instead of biological research, the end game was to use the rockets to take atmospheric samples following a series of high-altitude nuclear tests. China has a long history of shooting stuff into the sky, as it is credited with inventing fireworks in the second century BCE. It was the Mongols, however, and the tribes of the Middle East that brought these protorockets, and so gunpowder, to the West, where it was adapted immediately for warfare.

These early rocket trials with dogs in China led to the Shenzhou Program that put the first Chinese taikonaut, Yang Liwei, into orbit on October 15, 2003. To test the life-support systems of the spacecraft, the Chinese sent up *Shenzhou 2* in 2001 carrying a rabbit, a monkey, and a dog. In his memoirs, *The Nine Levels between Heaven and Earth,* Yang Liwei writes that not only has China flown dogs in space, but Chinese taikonauts eat dog in space. And not just any dog, but Huajiang dog from Guangdong Province in the south of China, touted for its nutritional benefits. In fact, popular belief in Huajiang is that local dog meat is better for one's health and strength than the super root ginseng. Despite eating dog, Yang Liwei notes, Chinese taikonauts eat "quite normal food" in space. An article in the UK's *Telegraph* reports that items on the menu aboard Chinese spacecraft include lotus root porridge, hairy crab with ginger, eel with green pepper, and baby cuttlefish casserole.

□

In early 1961 the Soviet team settled on two dog flights as a final test of the Vostok spacecraft before they sent a human being into space. "The aim was to test the entire Vostok system," said Yazdovsky in *Roads to Space*, "including the space suit, the ejection seat, and the life support facility." On March

9 a dog named Chernuska (Blackie) made one orbit with a wooden mannequin the team called Ivan Ivanovich. Also on the flight were forty gray mice, forty white mice, forty black mice, guinea pigs, reptiles, human blood, cancer cells, plant seeds, various microbes, and fermentation agents. American and British scientists called the flight "a veritable Noah's Ark," Yazdovsky said, "carrying all the species represented on Darwin's evolutionary scale." Ivan Ivanovich wore the orange SK-1 pressure suit that the first cosmonaut would wear, and some of the biological experiments—mice, guinea pigs, microorganisms—were stowed in his chest cavity and abdomen, his hips and thighs. He rode in the ejection seat to test that system because the first cosmonaut was going to bail out and come down under a parachute. Ivan Ivanovich was like Pinocchio: not quite a man but not quite a mannequin either. The little black space dog, Chernuska, rode with the remaining biological experiments in the pressurized capsule. Most of the space dogs were white or mostly white, so Chernuska was a rarity by her coloring, along with Mishka (Little Bear), who was killed on her second flight in 1951, and Malyshka (Little One) who flew in 1955 and 1956 and was recovered both times.

Burgess and Dubbs tell the story of one of the members of the medical staff, Dr. Abram Genin, who, against regulations, strapped his old Pobeda wristwatch to Chernushka's leg as he helped prep her for the flight. After graduating from the military academy, Genin received the watch as a gift and now wanted to get rid of it. He tried to break it with hard use—swimming with it in the sea, dropping it on the floor. The watch kept on ticking. As Chernushka was loaded into the capsule, he strapped it to her leg "hoping he'd never see it again," he said in a 1989 interview with the Smithsonian. Did he think the rocket might explode or the dog would be lost in the Siberian wilderness? Did he have so little confidence in the rocket and the Vostok spacecraft? In Chernushka?

Ivan Ivanovich ejected from the spacecraft and Chernushka came down in the capsule, both landing in Siberia far to the east of Baikonur. Yazdovsky led the search and recovery team with a general named Nikolai Kamanin. It was snowing, and the wind stirred the snow, reducing visibility. According to Burgess and Dubbs, the team flew into a remote town, then traveled by truck as far as they could go, tracking Chernushka's capsule. Somewhere along the route they acquired horses and rode in through hard country to the landing site near the town of Zainsk, Tatarstan. Locals had seen parachutes descending from the sky, then found a strange capsule on the ground, and farther off they could see what appeared to be a man in an orange flight suit lying unresponsive in a field. They wondered if he was a foreign spy and why he wasn't moving. Maybe he was dead. The rescue team arrived, ignored the man in the field, and saved the dog, which emerged from the capsule wearing that watch. The team held her up for the fascinated crowd to see, the dog that had just flown in space.

Later the team tracked the watch back to Genin and returned it to him. "He was still wearing the watch at the time of the interview in 1989," write Burgess and Dubbs, proving it was nearly indestructible.

Upon her death, Chernushka's body was stuffed and put on display in the museum at the Institute of Biomedical Problems in Moscow. In 2011 the schoolchildren of Zainsk held a contest to design a memorial to Chernushka, whose story was legendary in the town. "The resulting monument features the trajectory of a spaceship looping around the Earth," writes Turkina in *Soviet Space Dogs*, "with Chernushka's head juxtaposed against it, proudly gazing skyward."

Ivan Ivanovich made a second flight on March 25, 1961, this time with Zvezdochka (Little Star), a white ragamuffin of a dog, hardly a dog at all, with dark ears and a dark patch around her right eye. She was given her name by Gagarin, who was present at the launch and would soon be launched himself. This time the team wrote the word *maket* (dummy) on Ivan Ivanovich's forehead. The pair rode into orbit and made one revolution of the Earth, and on the way back Ivan Ivanovich ejected while Zvezdochka rode down in the capsule. Both were recovered safely.

○

In spring 1961 the young Soviet air force officer, Yuri Gagarin, became the first man in space, making one orbit of the Earth in a Vostok spacecraft. Flying high above the planet, he crossed over the United States, over Africa, and over miles of blue ocean. From the window of his spacecraft, he gazed on the splendor of the Earth. "It's beautiful," he later said. "What beauty!" On his descent, when Vostok reached about 23,000 feet altitude, Gagarin ejected as planned and rode down under an orange parachute. Hanging in the sky, he could see the great Volga River of his native land, a field camp in the countryside, and some women tending a calf. He landed on his feet in a plowed field near the town of Engels, about 700 miles north and east of the Baikonur Cosmodrome where he had started his journey, where Laika before him had started hers. Dragging his chute behind him, he walked to the top of a hill and saw a woman and a young girl approaching. Gagarin was still a man, but too, he was something more: a cosmonaut, and the very first. When they noticed him, the woman slowed and hesitated. Frightened, the girl ran away. "I'm one of yours, a Soviet," Gagarin yelled after them. "Don't be afraid." He walked up to the woman and explained that he had come from outer space, and he needed to find a telephone to call Moscow.

□

Gagarin is immensely important as the man who went first, but before him were the animals, the fruit flies and rabbits and cats, the fungi and fish and spiders, the chimpanzees, and the space dogs, with Laika as their ambassador. What is it one makes of these events? How are we to understand the many trials and accidents and sufferings of the animals flown into space, their sacrifices, the physical proof that they could withstand the rigors of a rocket launch, the reentry of a spacecraft, life in a confined capsule, the unknowns of microgravity, bombardment by high-energy cosmic particles. What does all this mean to us, for us, about us? What does it mean that we use animals for our own designs, our own purposes, to improve human life, for wealth and power? What does it mean that we sacrifice them instead of ourselves?

Animals advance us. Their fantastic achievements become

our achievements. Our civilization, on which rests the advancement of our technologies, from agriculture to computers to space-faring, would not be possible without animals. But we do not own the animals of the Earth. They are not here for us alone. They are beings in their own right, and this is how we should think of them. We use animals to learn, and we learn from animals, but they belong to themselves. It is as if the storehouse of human knowledge was given to us by the animals, and sometimes at great expense to them. When animals die in service to us, I think it takes something from us, some piece of our humanity, even while it reminds us that we are human. How do we live inside this contradiction, that the animals we love best—chiefly, the dog—we also sacrifice to the monument of civilization, to the monument of ourselves? If we cannot come to any clearer purpose than a stated contradiction, at the very least when we turn again to the animals for help, and that help is given, let us not forget where it came from. It was the animals, it was the space dogs, who taught the cosmonauts to fly.

❍

In 1960 the Soviet government gave Korolev a house in a forested park in north Moscow near the present-day Memorial Museum of Cosmonautics. He lived there until his death in 1966, and these were his most productive years. He is said to have worked sixteen to eighteen hours a day, spending some six months away from home each year. But when he was working in Moscow, he came home every day for lunch. He did not care for foods that required effort to eat, fish with bones, for example, a waste of time to him. He loved pickled herring. He was a reader, and like Tsiolkovsky, Goddard, and von Braun, he loved science fiction, especially Ivan Euremov's novel *Andromeda's Nebula* and the novels of H. G. Wells and Jules Verne. He kept the works of Leo Tolstoy among his prized books, and he could recite passages from *War and Peace*. Like the great novelist himself, Korolev kept a set of dumbbells in his study, believing in physical exercise to sharpen the mind.

Korolev was the kind of person other people orbited around: his second wife, a few rumored mistresses (for he had a great fondness for women, and they for him), his co-workers, the first cosmonauts, dignitaries and officials of the government. Korolev was very close to Gagarin, who often came to the house to be near him. The house was alive with such people, passing in and out, stopping by for a talk or coffee, a meal or a movie. Korolev kept a reel-to-reel projector in the front sitting room, rare in those days in the Soviet Union (or anywhere, for that matter), and he used any excuse to assemble a party. There in the sitting room he also kept a television. After his first heart attack in December 1960, he avoided the excitement of news and press events. During the great celebrations of Gagarin's successful flight into orbit in 1961, Korolev stayed at home to watch it on TV.

Korolev had great taste in art, or so I was told by my guide when I visited his Moscow home. Hung on the walls are several paintings he acquired not long before his death, evidence of a premonition, my guide said. I asked after the titles and the names of the artists. One work, entitled simply *Landscape*, was touted as the work of a French painter of some renown, but my guide could not name him. Two other paintings, both set at dusk, are entitled *Evening Landscape* and *Evening Landscape by the River*. For both, the painters were unknown. The mood of these paintings is dark, to be sure, a setting sun and a foreground of failing light. Fitting scenes for a man fixated on vanishing.

At the top of the stairs before the entrance to his study is a framed map of the moon, perhaps six feet by six feet in size, a handcrafted original. It was given to Korolev by a scientist in Saint Petersburg to honor his dream of planetary exploration. Korolev believed the moon to be rocky and solid, that a spacecraft could land on it and a cosmonaut could walk on it. Others during this time believed the moon to be gaseous, like the gas giants Jupiter and Saturn. The first spacecraft to land on the moon, *Luna 2*, which Korolev's team launched in 1959, proved he was right. Further into the study is a globe, a gift from the rocket engineer Valentin Glushko, who is said to have sent Korolev to prison but afterward became his boss (for a time), his colleague, and his friend. On the globe's stand, Glushko has written: "My dear friend—I wish you to see the Earth like this from space."

Korolev's health problems, which began during his imprisonment, were troubled by his impossible work schedule and unremitting stress. He suffered from a form of kidney disease, cardiac arrhythmia, an inflamed gallbladder, hearing loss (likely from test-firing rocket engines), and intestinal bleeding. On the outside people saw a hale and powerful man who was at the extreme edge of human ingenuity and industry, but his body was failing. In his book *Korolev*, James Harford cites a letter Korolev wrote to his second wife, Nina Ivanovna, during one of his stays at the launch facility at Baikonur: "I am unusually deeply tired, and sometimes the little heart aches a bit." Korolev dared not take any rest, however, for he was convinced that without him to drive the newly formed space program, Khrushchev would pull the funding and cancel it. So he worked ever harder, despite knowing he should not. "I can't work like this any longer," he once said to his wife.

To correct a bleeding polyp in his large intestine, on January 5, 1966, Korolev checked into the Kremlin hospital, a facility catering only to top Soviet state and Communist Party officials. The long black wool overcoat and shoes he wore that day are in the closet near the front door of his house. In the hospital, Korolev underwent a series of tests and then into surgery on January 14. In his book, Harford offers the details of Korolev's death as told by his daughter, Natasha. During the routine surgery, Korolev began to bleed, a persistent bleeding that required his surgeons to cut into his abdomen. There they found a cancerous tumor and went to work to remove it. Korolev was under an anesthetic mask for eight hours, and after the surgery he never regained consciousness. Had the surgeons intubated him, Harford suggests, he might have made it, but his broken jaw from the torture he endured in the gulag had not healed properly and prevented the tube from going in. Still, the tumor was malignant, and according to Harford, "Korolev would not have lived more than a few months, even if he had not been operated on." His body was cremated and his ashes interred with honors in the Kremlin Wall. He was fifty-nine years old.

Inside the front door of Korolev's Moscow home, at the foot of the stairs on a table near the telephone, is a sculpture, *To the Stars*, a replica of a larger original that is mounted on the grounds of the Theater of the Soviet army in Moscow. The first three cosmonauts in space (Gagarin, Titov, and Nikolayev) presented the sculpture to Korolev as a gift. It features the signatures of all eleven cosmonauts who flew in space during Korolev's lifetime. The sculpture is of Prometheus, his left arm outstretched, releasing a rocket to the stars. He leans dramatically forward, his body bare but for a cloth draped around his waist that flows and snakes about him. The rocket is to go ever upward, and Prometheus has become part of the rocket, launching into the cosmic void on a voyage of discovery and adventure. Behind the sculpture is a staircase leading up to Korolev's study. The staircase rises a few steps to a landing, where it turns and rises again past a bright and sunny window looking out into the forested park beyond the house. Korolev would often sit on the second step from the top and look out the window through the leaves of the trees. It was here, my guide told me, that Korolev had his best ideas.

❍

In 1966, a month after Korolev's death, the Soviets launched a final space dog mission. Both the US and the USSR had turned their sights on landing men on the moon, and this flight would test hardware and life-support systems on an extended stay in space. Ugolyok (Little Piece of Coal or Ember) and Veterok (Little Wind), along with other biological experiments, were sent into orbit on *Kosmos 110* for a twenty-five-day mission. The dogs would be pushed into an elliptical orbit with an apogee of 560 miles and pass through the lower Van Allen Belt, where radiation levels were measured between six and twenty-five times higher than on flights in lower orbits. In addition to gathering data on the effects of microgravity on the body during a long flight, the team would also come to better understand the effects of exposure to such high doses of radiation.

Ugolyok was a fluffy dog, dark as coal as her name indicates, and bearing a rounded mane about her face, ruffed out like a male lion. She was tall and handsome, a dog of appealing conformation, a dog you might want to take home. Veterok, with her short legs and shorter hair, her ears bent over in most photographs, appears as a kind of sidekick, a dog you would keep only if you had another. For that, they were suitable companions. Video footage of the two dogs on a walk (the caregiver, a woman with a beehive hairdo and wearing a white lab coat and black heels) shows them to be energetic, playful, dogs at the peak of their youth. Turkina notes that both dogs had other names before their flight. Ugolyok had been called Snezhok, which means Snowball, an irony, since she was almost all black. And Verterok had been called Bzdunok, which means Little Fart, perhaps a commentary on her personality or maybe her behavior. Despite the name change, someone, it seems, thought to preserve the humor, as Veterok's name changed from "little fart" to "little wind."

During the flight the dogs would take food and water through stomach tubes. Veterok would be administered doses of a new antiradiation serum through an intravenous needle, while Ugolyok would not. The team could then compare the condition of the two dogs when they returned to Earth. If the serum worked, it might be useful in treating cosmonauts on a mission to the moon or possibly for radiation sickness on Earth, a salve for the threat of nuclear war.

An article in *Time* published a couple weeks after the flight suggests that Ugolyok and Veterok were "moon dogs," the "immediate predecessors of the moon dogs the Russians have said they intend to send into lunar orbit ahead of man." Such speculation is corroborated by cosmonaut Gherman Titov (the second human in orbit), when he predicted with some disappointment that dogs would land on the moon before humans. Dogs had become so practiced in spaceflight that the newest space race was not between Soviets and Americans, it seemed, but between humans and dogs.

After twenty days in orbit the team discussed bringing Ugolyok and Veterok back early. The air quality in the capsule was still acceptable, but it was in a state of slow and steady decline. According to Asif Siddiqi in *Challenge to Apollo*, a landing commission of twenty-five members discussed the issue throughout the night. Yes, they agreed, if the dogs were going to survive, they had to be brought home now. The dogs landed near Saratov, Russia, after twenty-two days in orbit.

Back on the ground, Ugolyok and Veterok emerged from the spacecraft alive but exhausted and disoriented. A later report, writes Siddiqi, notes that the two dogs had lost 30 percent of their body weight and showed signs of "muscular reduction, dehydration, calcium loss, and confusion in readjusting to walking." According to Turkina, both dogs recovered fully after about ten days and both gave birth to litters of puppies. A Russian writer has claimed that both lost all their hair after returning to Earth, but this claim cannot be substantiated. Whatever became of Ugolyok and Veterok in ensuing years may be less important than their achievement: they set the duration record for space dogs, one that will not be broken until, maybe one day, we take our dogs with us to Mars.

FIVE

◇

A Face in the Window

I require a You to become; becoming I, I say You.

MARTIN BUBER
I and Thou, 1923

Preparations for Laika's flight began in Moscow at the Institute of Aviation Medicine where she lived and trained. During its ascent and after the satellite entered orbit, the telemetry system on the spent booster would track Laika's blood pressure, heart rate, respiration, and movements and send that information to a ground station. To measure a dog's blood pressure remotely required a blood pressure cuff on a timer that Laika wore around her neck. Anyone who has ever walked a dog on a leash knows that its head and neck are powerfully muscled. That, combined with a good coat of protective fur, makes it nearly impossible for a blood pressure cuff to depress a dog's carotid artery and so measure its blood pressure. To solve this problem, Gazenko and Yazdovsky performed surgery on Laika to draw her carotid artery out and sew it into a flap of skin close to the surface where the blood pressure cuff could make contact. It was a delicate operation, and it required a good ten days to heal.

To record Laika's heart rate while she was in orbit, Gazenko and Yazdovsky surgically implanted two silver electrode rings, not more than two-tenths of an inch in diameter, beneath the skin on her chest. To these they attached wires and drew those wires beneath her skin up to the top of her back near her shoulders, one on each side. Laika must have looked like a satellite herself, a little sputnik, with those long antenna-like wires emerging from her back. Albina too underwent both of these surgical procedures, because if something happened to Laika in the final moments before launch, Albina would have to fly.

Equipment to record Laika's movements and respiration did not require surgery. The harness Laika wore around her chest included a gauge that measured the inflation and deflation of her lungs. Her movements were measured by a wire, this one attached to the outside of her harness and wound onto a spindle drum controlled by a spring at the rear of the capsule. When Laika moved away from the drum, she drew the wire out. When she moved toward the drum, the spring engaged and wound up the slack. A sensor recorded the length of the wire, drawn out or wound up, and the ground crew could then determine if she was moving about and roughly where she was: pressed up against the back of the capsule, somewhere in the middle, or far forward, close to the window.

◻

Not long before Laika was flown to Baikonur for launch, Vladimir Yazdovsky took her to his Moscow home to play with his children, because before he was a scientist, or a Soviet, or even a Russian, he was a human being. And human beings are inseparable from dogs. "Laika was a wonderful dog...quiet and very placid," Yazdovsky writes in his memoirs. "Before her flight, I brought Laika home and showed her to the kids. They were fascinated by her behavior and her beauty. They played with her and pet her. I wanted to do something nice for the dog since she didn't have much longer to live."

It was early in winter, and the Moscow night would have been cold with temperatures near freezing. At two years old, Laika was still a young dog, and if she was quiet and placid, she may have entered the house with a slight hesitation mixed with the excitement of seeing so many new people, young people, moreover, the children filled with energy and excitement themselves. What did Laika do during her time in Yazdovsky's home? How long did she stay? Did the children beg for her to sleep in their rooms? We simply do not know.

Among the stories told about the space dogs of the Soviet program, this is one of the most important. It speaks to Yazdovsky's humanity, and to the humanity of the entire Soviet team. It tells us that these scientists and engineers cared deeply about these dogs—they loved them—and treated them as friends and colleagues, as working dogs. It tells us that they felt not guilt, I think, but empathy for Laika and her mission. Laika was to be sacrificed, and while I found no record of anyone on the team in real opposition to that sacrifice, there is plenty of evidence that the people who sent her into space did so with a heavy heart.

Yazdovsky was in a leadership position, and he had great

authority over the dogs' training and care. And yet the dogs did not belong to him, and they did not belong to his superior, Korolev. Beyond the fact that the space dogs belonged to themselves (but this is another kind of truth), they belonged to the Soviet Union, to the mission that was the Soviet Union, and now, I think, these many years later, they belong to all of us. If Yazdovsky was able to take Laika home with him, it was a great feat indeed. The Soviet rocket and missile program was as secret as it was successful, and to remove from the kennels the one dog that had been so carefully trained for this historic flight would have been risky and in defiance of regulations. Yet the story persists, substantiated by Yazdovsky himself. When I related the story to Sergei Khrushchev, he admitted he had not heard it before. I asked if he thought it were true. "Taking the dog home?" Khrushchev said. "I don't think this was possible. How you receive permission to do it? It isn't reasonable. But then again," he said, taking a long breath, "it's Russia. And everything can happen in Russia."

▢

The space dogs of Soviet Russia were not lab animals, I think. They were cosmonauts, highly trained working dogs with a job to do. While their work was dangerous, and some of them died doing it, so is the work of war dogs, police dogs, search-and-rescue dogs, herding dogs. A dog does not choose its work, but rather it is bred and trained to perform the work it does, the work we need it to do. And work, as the original agreement between humankind and dogs, is the underpinning of the bond we form with them.

hard to say, but these dogs were real professionals. They submitted themselves to training, and perhaps when the sensors were fitted on them, they understood something serious was going on." Oleg Gazenko agrees: "No one working on the experiments involving animals saw them as just dogs," he said in *Space Dogs*. "We saw them, rather, as our colleagues, as friends. It was amazing how, even during the sometimes painful procedures, when some medicine had to be injected, or some hair had to be removed so we could attach the sensors, the dogs never took it as an act of aggression or unfriendliness. On the contrary, they would turn and give you a lick on the cheek."

It is unhelpful, I think, to regard the space dogs as victims, nor can we think of them as choosing their life among the stars. So what are we left with, when it is so painful to imagine these dogs enduring the stresses of training and then of spaceflight, and some of them dying upon impact with the Earth, or dying in a fiery explosion, and some of them dying in space? We are left with an emptiness that arrives with the fullness and necessity of human endeavor in which dogs, and other animals too, are our companions, our subjects, and sometimes our sacrifices. We love them, and we sacrifice them anyway.

I support the rights of animals to live as they evolved to live, and all animals live in relationship with other animals, humans included. A distinction may be made here between working animals that are bred and trained to work, want to work, even need to work, and animal research, the use of animals for experimentation in scientific laboratories. Much of the opposition centers on animal research, as opposed to using animals for work, but both are unsolvable problems, even as the universe, it seems, allows for such problems. Not every thesis has an antithesis. I do not here wish to take up a position for or against animal research but rather to acknowledge that humankind has benefited greatly from this relationship, including those people who profess to be intolerant, who work toward a world in which no animal is harmed by another. Those people have benefited too.

In *Billions and Billions,* famed astronomer Carl Sagan writes about his struggle with myelodysplasia, a disease of the bone marrow. After his diagnosis, his doctors told him he did not have long to live unless he underwent a bone marrow transplant. Sagan struggled with the fact that animal research is responsible for the development of this procedure. "In my writings, I have tried to show how closely related we are to other animals," writes Sagan, "how cruel it is to inflict pain on them, and how morally bankrupt it is to slaughter them to, say, manufacture lipstick. But still, as Dr. [E. Donnell] Thomas put it in his Nobel Prize lecture, 'The marrow grafting could not have reached clinical application without animal research, first in in-bred rodents and then in out-bred species, particularly the dog.'" The disease eventually took Sagan's life, and to the end he remained deeply conflicted by benefiting from treatments that relied on animal research. For him, as for many of us, it is an unsolvable problem.

The space dogs flew in the days before the standardization of laws governing the care and treatment of animals, and yet in my research I only found evidence that the team under Korolev and Yazdovsky exemplified what we today hold as the code of the animal researcher:

Refine—research with as little pain and trauma as possible

Reduce—limit the frequency and numbers of animals used in research

Replace—remove the need for animals in research by finding other options

In remarks he made in 2005, the Dalai Lama offered his position regarding animal research, echoing the code of the animal researcher: "I encourage the minimum use of experiments on animals, the absolute minimum amount of pain. Only perform highly necessary experiments, and as little pain as possible. If it must be done, [if that is your path, it is compassionate] to kill out of necessity, but only with empathy. Hold in you the sense of the compassionate: 'I [acknowledge] that I exploit this animal to bring greater benefit to a great number of sentient beings.' You must feel the sacrifice, in your heart. It is never made lightly."

◌

Yazdovsky, Gazenko, and Abram Genin (the man who would later strap his watch to Chernuska), along with the three dogs—Laika, Albina, and Mukha—boarded a Tupolev TU-104 to fly to Tashkent, and from there on to Baikonur on a smaller Ilyushin-14 prop plane for the launch of *Sputnik II*. The Tupolev TU-104 was a Russian turbojet aircraft with twin engines, one of the first turbojets in operation in the world. It was too big, really, to land on the short runway at Tashkent. It *would* land at Tashkent, but not without causing Korolev a great deal of worry and stress. And his worry was justified, as on that single plane flew the entire team that was to make *Sputnik II* possible. If it crashed, a great deal would be lost, maybe everything. Korolev spoke to the pilot before the flight: *remember,* he said, *you are transporting the* haut monde *of Soviet science and engineering. You must bring them in safely.*

At Baikonur, the team was making final preparations, still racing the clock to launch *Sputnik II* in time to celebrate the Revolution on November 7. For days now, Radio Moscow and the Soviet press had been updating the country about the location of *Sputnik I* on its journey around the Earth. Everywhere in its path, people watched the skies, hoping to see it pass overhead, and listened to radio broadcasts to hear that haunting sound: "beep, beep, beep, beep." The press also spoke of a satellite nearing completion, a new satellite that would be even more spectacular than the first because it would be carrying a live animal into space. On October 27, before coming to Baikonur, Laika was "interviewed" on Radio Moscow and, as if on cue, she barked.

With Laika on site at Baikonur, along with her handlers and medical staff, *Sputnik II* could be prepped and loaded for launch. One night Baikonur's deputy commander, Anatoly Kirillov, was watching the three little dogs running about the control room. He and his co-workers felt sorry for the "little mongrels," as he calls them in *Roads to Space*. "We lamented the fact that one of them would soon die a gruesome death in orbit." His lamentation was interrupted, however, when someone came rushing into the control room to announce that *Sputnik I* was about to pass overhead. Korolev, members of the Soviet government, and other personnel assembled outside under the night sky. According to Burgess and Dubbs, people all over the world had been watching *Sputnik I* pass overhead, but the team that built the satellite had never seen it. At Baikonur, the satellite would be visible only briefly and just above the horizon. It would be difficult to see, but on this night, even if only for a moment, the team had a chance. "When the satellite did appear," said Kirillov, "it rose high in the sky as it moved from the south-west toward the north-east. It kept us spellbound for several minutes until it finally vanished."

Ivanovsky also records this moment in *The First Steps*. "We were staring at the horizon," he writes. "Minutes passed by. . . it was not so important whether we were going to see [*Sputnik I*] or not, it was essential that it was flying high in outer space, and it was an established fact! A few minutes later someone spotted a moving point of light." There it was, *Sputnik I*. It looked to Ivanovsky like a firefly aglow on a summer evening. "The firefly seemed to move proudly, confidently, and even at ease," he writes. "Many people wiped off a tear. I have seen, more than once, the twinkling firefly in the sky, but that first sighting would be stored in my memory forever."

Most of the assemblage returned to their workstations or to their leisure, Kirillov reports in *Roads to Space*, but he and a few others lingered in the cold with Korolev. The chief designer stood in silence, his breath visible in the air, gazing up at the night sky as if the satellite might appear again. "Well, my friends," Korolev finally said, "I've known for a long time that satellites could be launched with the help of this." He put his finger to his temple to indicate the human mind. "But that we actually managed to pull it off using our own heads and hands and this amazing thing," Korolev then motioned to his heart, "I find absolutely incredible."

◌

On the morning of October 31 one of the caregivers took Laika out for her morning walk. This was the usual routine. Perhaps they even walked a familiar route, one they had been walking since the day Laika arrived at Baikonur. For Laika, then, there was nothing extraordinary about this walk, nothing special about this day.

At 10 a.m. Laika was taken to the medical facility for flight preparations. "She was quietly lying on a shining white table," Ivanovsky writes. "The technicians cleaned her skin with a weak solution of alcohol, carefully combed her hair, and applied iodine water and streptocide powder to the spots where electrodes were implanted under the skin to record the ECG. These procedures took two hours." Two hours is a long time for a young dog like Laika to allow the team to work on her, and yet she did, so patient and amiable was her character. The procedures were familiar to her too, part of her training. The iodine water and streptocide may have soothed her itchy incisions, now mostly healed, where her carotid artery lay in a flap of skin, where the electrode rings were seated beneath the skin in her chest, where the lead wires emerged from beneath the skin high on her shoulders. The combing, those long, gentle sweeps of the brush and the comber's hand must have also soothed her. They cared for this dog, even as a mortician might care for the dead, might care for the body soon to be laid to rest, already to be mourned.

It was noon before they finished. The next step was to dress Laika in her vest and harness and attach the waste collection bag. Ivanovsky reports that at this point, Korolev entered the lab dressed in a white lab coat. He examined Laika and the preparations the team had made, all the while speaking softly to her, placing his hands on her body, ruffing her about the neck and scratching her behind the ears. Korolev would remain with Laika until she was ready to be sealed inside her capsule. He attended to each detail of her preparation, checking the work of the team to be sure that once Laika was in orbit the scientists on the ground would receive the data they needed, and to be sure that Laika received whatever she needed.

The team prepped her capsule, loading the space dog food into the feeding tin, the chemicals in the regeneration unit that would help scrub CO_2 from the air, purifying it to make breathing possible for as long as possible. At 2 p.m. Laika went into her capsule, and her sensors were attached to the recording equipment. All this was familiar to her. They sealed the door. Now the only connection Laika would have with the outside world was that little round window. The people who had cared for her and trained her could see in, and she could see out. She waited patiently inside the capsule while the team made a series of tests on the equipment, the equipment measuring Laika's heart rate, blood pressure, her respiration and movements. The oxygen levels inside the capsule were adequate, and the CO_2 scrubbers were working just fine. Faces peered in at her through the glass. *There she is,* someone might have said. *She's doing just fine. Everything seems to be working just fine.*

At 1 a.m. on November 1, "the [capsule] with Laika inside was delivered to be put into the rocket," writes Ivanovsky. "A little 'dog house' was slowly lifted on a large crane hook. The assemblers picked it up with caring hands and secured it in place." When the team moved her from the preparation room, she would have seen the bright lights of the room turn to a black night sky through her window and then felt the lift and perhaps twirl of her cabin as she swung from the crane, drawn up to the top of the rocket, and moved into place.

"Of course, we knew this satellite would not come back to Earth," writes Adil Ravgatovna Kotovskaya in her article "Why Were Flying Dogs Needed for Rockets and Satellites to Launch Yuri Gagarin?" "Therefore, we saw Laika off saying goodbye to her and asking for her forgiveness."

◇

Inside the capsule Laika waited, and she would continue to wait for three days as final preparations were made for launch. The team monitored her vital signs and observed her through the window. Her respiration rate was at 16 to 37 breaths per minute, her heart rate between 68 and 120 beats per minute, all within the normal ranges the team had recorded during her training. Through the window, Burgess and Dubbs write, Laika "could be observed...sleeping, feeding, or reacting to a human face peering in at her."

For Laika this was just another training exercise, like the

many she had been through before. In her experience, when she went into the capsule, she always came out, even if a number of hours passed, or a number of days. Why would this time be different? A dog trainer once told me that a dog becomes accustomed to its people leaving the house and returning. When you leave the house, your dog knows, in the way a dog can know, that you are going to return because you always have. I imagine Laika was content inside the capsule because it was familiar, she had trained for this, and she knew that someone, Gazenko or Yazdovsky or one of the caregivers, would release her. She would again be leashed and go outdoors for a walk to stretch her legs. She would again be reunited with the other space dogs she lived and worked and walked with. She would return to her kennel, her familiar bed and water dish, the familiar dogs on either side of her, the familiar lights and sounds of the kennel, its smells. She would again be fed at the regular time. She would again go home. For Laika, the capsule was a kind of home too, one that she had grown accustomed to in her training. She was comfortable inside the capsule and made to feel comfortable, well attended, looked after. For Laika this was a day like any other.

☐

A dog is a social animal with a face that asks for your attention, your acknowledgment, your love. The dog looks at you, and through its dark, often sad-looking eyes it asks to be recognized. I am alive, the dog is saying, and you are alive, so let's acknowledge each other. We feel something penetrate us to the center in looking into the face of a dog, something vital and fundamental, something that was there before we were here, distinctly human and yet resident in the kingdom of animals. When we look into a dog's face, we want to help, somehow. "The face opens the primordial discourse whose first word is obligation," writes French philosopher Emmanuel Levinas in *Totality and Infinity*. Here Levinas is writing about human-to-human relationships, but I think we can extend his sensibilities to human-to-dog relationships, because the dog is a member of the human community, a resident of hearth and home, because humans and dogs have been together for a long time, for as long as civilization. Dogs and humans built civilization together. This obligation Levinas writes about, this compulsion to act or to help another human being or a dog, tugs at us until we reject it or accept it, either of which defines the boundary of the relationship.

A relationship with a dog does not encourage possession, however. "The face resists possession," writes Levinas, "resists my powers." The face "is present in its refusal to be contained. In this sense it cannot be comprehended, that is, encompassed." What is at stake in looking into the face of a dog, in coming into relationship with it, is the fragile beauty of trust. If a relationship is to develop, it must be based in trust, and at its edge is a necessary moment of determination: friend or foe? The dog asks this of us, and we ask it of the dog. And the face makes this determination possible. In the dog's face "is the primordial expression," writes Levinas, "is the first word: 'you shall not commit murder.'" You shall not murder me, the dog says, which means, you shall not possess me. Only after this limit is established does a relationship of trust become possible.

In his essay "Why Look at Animals?" John Berger affirms that this is so. When an animal looks at you, Berger writes, you "become aware of [yourself] in returning that look. And this exchange sets up the condition for our parallel lives." We need each other—animals and humans, dogs and humans—but as much as our lives are entwined, they remain parallel too. The look we exchange with animals, writes Berger, crosses a "narrow abyss of noncomprehension," so that this awareness of ourselves that is mirrored in the face of an animal, with our shared gaze, does not necessarily include an understanding of that animal. Our lives are always separated by a space we cannot cross, a space across which we fail at understanding. In considering more specifically the space between humans and dogs, we may become adept at reading what a dog wants (food, a walk, to be in our presence), but we cannot know what a dog feels or even comprehend a dog's feelings. We can share something with a dog, but we can't presume to understand it. This condition, this state of noncomprehension, is fine by us. We don't need comprehension to love a dog, or any other animal for that matter, because dogs offer us something we can't get anywhere else. "With their parallel lives," Berger goes on to say, "animals offer man a companionship which is different from any offered by human exchange. Different because it is a companionship offered to the loneliness of man as a species."

Companionship with animals, especially with dogs, I think, is an antidote to loneliness. It is a gift that animals may offer to us, a gift that dogs may offer, and they give it freely, of their own choice and will.

As the scientists and engineers looked in at Laika through

her window, they saw her dark eyes, the drooping tips of her ears, her mouth open as she panted lightly. As the scientists and engineers looked in, what they saw was Laika looking back at them, Laika reaching out with her gaze to make contact, to create something between her and them—trust, perhaps—and thereby define the stable ground all relationships are built on. The scientists and engineers would have seen something else too, something in the margins of the glass and in its center, shifting as the light shifted with the shadows they themselves cast over Laika inside her capsule. They saw their own reflections, their own positions in space and time. They could not have but marked that moment as they noticed themselves noticing her, noticing in her a concordant innocence for all her unknowing about where she was going and what was going to happen to her. In that moment they must have acknowledged, even if only privately, that Laika was both simultaneously alive and dead—a kind of Schrödinger's cat— that the capsule was a container inside which she was alive, but from which she would never escape. In that moment the window became a medium through which they might witness the drama of her end, if only they could follow her that far, and a medium through which they might imagine their own ends and the fate of our own capsule, the Earth, which will not last forever. Where Laika was going—into the stars, into death—the scientists and engineers knew they were going too. Like her, they would go alone, as we all must go one day. Through that window, then, we see Laika's face, and we feel triumphant and elevated in her company, and also impossibly lost and alone with or without her, each of us alone in the cos-

send a man into orbit. The only solution, as Yazdovsky and Alexander Dmitrievich saw it, was to open the capsule, normalize the air pressure, and reseal it. So close to launch, this was an odd request, and it might cause a delay, but the investment in this mission was immense—time, money, resources, engineering and political capital, Laika's life—and no one wanted to miss this opportunity for good science. Korolev granted permission to make the adjustment. In the sources I read, however, there is some question about the authenticity of this permission from Korolev. It is possible that Yazdovsky and Alexander Dmitrievich were denied permission or didn't ask for it at all. But if either was the case, that information would likely have come out after the launch, and the two men disciplined. I found no such record.

Laika's capsule had been designed with a breathing opening on the outer shell sealed by a screw cap. Removing that cap was the easiest way to equalize the pressure. Yazdovsky and Alexander Dmitrievich explained the situation to the engineers on site, one of whom was Ivanovsky. It may have required a bit of coaxing and reassurance to convince them that Korolev backed this strange request, that there was time before launch to make this happen. That done, everyone stood by as one of the engineers unscrewed the cap. The pressure now normalized, the cap could go back on, but Yazdovsky and Alexander Dmitrievich had something else in mind.

"They literally attacked me, especially Alexander Dmitrievich," Ivanovsky writes in *The First Steps*. "Please, I beg you," insisted Alexander Dmitrievich, "Let's water Laika!"

Ivanovsky calls this moment in Laika's story "a trickery,"

and Burgess and Dubbs describe it as "a subterfuge." Whatever the sneakiness of the plan, after three days inside the capsule Laika needed water. She had eaten all of her space dog food, and while it would have helped with her thirst, it was not enough. She was probably already suffering from dehydration, her body's systems slowing down. Without water, Laika was not going to live much longer, and she had not even left the ground. "Frankly speaking," Ivanovsky writes, "we all were willing to comfort Laika's life in space a little bit."

In some re-creations of this moment in drawings, books, and animated shorts, the team opens Laika's window to give her water, but the window was not fixed on a hinge that could be opened. It was sealed tight, all the way around. The capsule too was sealed. It did not include a door but was rather a cylinder with an end-cap, and the end-cap contained the window. If either the window or the end-cap had to be removed to equalize the air pressure, it would have been impossible to complete the job and still launch on time. Korolev would not have given permission for such a delay. The image of Laika with her window open wide, the medical staff and engineers comforting her, giving her water to drink from a bowl, patting her on the head, is a comfort to us, but it is a fiction. The reality was far more clinical.

Alexander Dmitrievich pushed a rubber tube onto the end of a large syringe and filled the syringe with water. In her condition, Laika was probably listless and groggy, perhaps mostly lying with her head down between her front paws stretched out before her, her ears tipped over on the ends. All the activity outside her capsule might have brought her to attention.

All the talk and bustle. What was going on out there? I imagine that she could see people moving through the window, and some of them were people she knew, who had fed her, walked her, trained her. And there was Yazdovsky, the man who took her home. "When Laika saw a familiar face through the window," writes Ivanovsky, "she showed all the signs of canine joy." What does canine joy look like? Perhaps Laika stood up inside her capsule, the restraining chains restricting her movement. We do not really know. Perhaps she wagged her tail, panted hard at the window, steaming it with her breath, her body wiggling now with the wagging of her tail, her head going back and forth too, her ears moving up and down in her wiggling. And perhaps she barked.

Alexander Dmitrievich dropped the tube through the breathing opening and filled the empty food tray with water. Laika drank. She drank it all and would have taken more, but the opening had to be resealed. *Sputnik II* was soon to launch. These were the last moments of Laika's life in human company. The men looked in at her, and she looked out at them. Then, Ivanovsky writes, she "gave a grateful nod led by her wet nose."

SIX

○

First Around the Earth

It made her think of Laika, the dog. The man-made satellite streaking soundlessly across the blackness of outer space. The dark, lustrous eyes of the dog gazing out the tiny window. In the infinite loneliness of space, what could the dog possibly be looking at?

HARUKI MURAKAMI
Sputnik Sweetheart, 2001

On the morning of November 3, 1957, Ivanovsky looked out from the observation site at Baikonur to see the rocket that would carry Laika into space sitting on the launchpad, a "bright, white rocket, like a candle, standing out against the cloudless and blue November sky," he writes. Everyone was there watching. Korolev was there, dignitaries and government officials from Moscow, scientists and engineers. Some held binoculars to bring the rocket in closer. The loudspeakers crackled, and a voice gave the ten-minute alert. Then one minute. And then a series of instructions, followed by a final command: "Launch!"

The engines fired, a great plume of smoke and dust and fire spreading out like a flower against the ground. With the engines ablaze, the rocket seemed to sit on the pad a moment, a moment more, and then it rose smoothly into a space above the pad as if hovering there, not sure if it should stay or go, and then it went, accelerating as it rose, faster and faster, that bright burning flare of the engines running beneath it, bright as the sun. The rocket made a long bright line high above and to the east, Ivanovsky reports, and as the first-stage booster separated "beautiful ripples appeared in the sky," ripples that could be seen as far downrange as Alma-Ata, just north of the great Issyk-Kul Lake in present-day Kyrgyzstan. Those ripples dissipated, and the rocket kept going in its speed, so high and away that it was beyond where anyone could see, beyond the Karman line, beyond the Earth and on into orbit, into that slip of space between where everyone is and where no one had yet to go.

"Laika flew off," Ivanovsky writes. "We rushed to the cars and headed to the telemetry stations where radio signals of Laika's heartbeat would be received, as if by an invisible wire. Was she alive? Did she survive the take-off, the g-forces, the vibration?"

"It was a historic event," said Alexander Seryapin in *Space Dogs*. "Standing seven kilometers from the launch pad, you can feel the Earth vibrating. And then the sound of the engines. You know it gives you the shudders." As he stood watching the rocket, tears came into Seryapin's eyes imagining Laika in her capsule. "I thought it was just me who felt that way," he said. "For the first time, I saw men really weep when they were told that Laika had gone into orbit."

Inside the telemetry station, Alexander Dmitrievich studied the information coming in from *Sputnik II*. Then he raced to the door and threw it open, nearly tumbling out. Ivanovsky was there and some others were approaching, obviously in want of the news. He "threw the thumbs-up to us," Ivanovsky writes, "all right! It was a victory. Laika was alive! She was whizzing above the earth, unaware of what was happening to her and where she actually was." The telemetry signals from Laika, the fact that her heart was beating, her lungs were drawing breath, she was moving about a little, meant that the team had proven "that it was possible to live in space, in the mysterious and unexplored world!" Ivanovsky writes.

More information came in. During launch, Laika's heart rate increased to more than 260 beats per minute, about three times normal for a dog. Her respiration increased too, to about five times normal. When the satellite entered orbit and microgravity, Laika's heart rate began to slow, along with her breathing, and then stabilized. She stabilized, sitting in her capsule, calmer now, the capsule familiar to her but for the strange sensation of microgravity. She was not floating about the cabin, as the restraining chains and the small space she occupied prevented that, but she would have sensed the immense g-force pushing down on her during the flight, and then suddenly nothing. According to the medical team on the ground, Laika had endured the launch well, and now she was up there, shooting around the planet at 17,500 miles per hour in an elliptical orbit with a 140-mile perigee and a 1,039-mile apogee. At this speed and orbit, Laika made one revolution around the Earth in about 103 minutes. She was now the fastest dog that ever lived.

The space dogs that had gone beyond the Karman line before Laika had lingered in microgravity for only a few min-

gravity. That 600 pounds is reduced suddenly and dramatically, but it is not reduced to zero. The term "zero gravity" is misleading because there is plenty of gravity in orbit. If there were no gravity in orbit, there would be no orbit at all. In zero gravity, a satellite or the ISS would simply fly off into the great unknown, the moon would not be the Earth's constant companion, and the solar system, even the galaxy itself, would not hold together. Gravity on a spacecraft in orbit is about 88.8 percent of what it is on Earth. The reason astronauts and objects float in Earth orbit is not because there is no gravity. They float because the spacecraft and everything in it is falling around the Earth at the same velocity. To be in orbit is to be in a continuous state of free fall, the spacecraft and everything in it, falling back to Earth. But in orbit a spacecraft's velocity is in equilibrium with gravity so that it will never reach the surface. The spacecraft falls around the Earth, and it just keeps falling.

Human astronauts ride into space positioned on their back because it's easier to breathe. They can inflate their lungs upward into the empty space above them. Lying on their bellies doesn't work, because under the force of 3 or 4g, astronauts would have to inflate their lungs by pushing the now tremendous weight of their body upward, again and again, for eight minutes. It would be like trying to breathe while lying on your belly with two or three people sitting on top of you. An astronaut's blood can flow almost normally while positioned on his or her back, as the force of acceleration is acting laterally on the body. If the body is vertical to the force of acceleration (toe to head, or head to toe), blood either rushes away from the brain, which can cause a loss of consciousness, or it rushes to the brain, which can cause cerebral hemorrhaging and possibly death.

In *The First Steps*, Ivanovsky describes film footage of rats riding into space on a rocket. "While taking off and accelerating, rats slow down, their legs are wide apart, their heads go lower and lower, and finally hit the floor," he writes. "The g-force presses an animal down to the floor, and it stops moving, its muscles cannot cope with the increased weight." He goes on to say that a "few seconds later the animal suddenly takes off from the floor and for a moment hangs somewhere in the middle of the box. No support! The rat begins to move randomly in the box. It goes from turning around its axis to flying into the corner, from spinning, like a spindle, to somersaulting." As weight is a primary indicator of up and down for all terrestrial animals on Earth, the rats lose "the sense of up and down, and [have] no points of support and no signals from legs and tail are being delivered. Only the vision continues on normally, but at first [they fail] to deal with the chaos of other perceptions."

Laika wore a flight suit to keep her sensors in place, but she did not wear a space suit. Her sealed capsule inside *Sputnik II*, however, helped minimized the noise and vibration from the rocket. These sensations were familiar to her too, as she had trained to tolerate them. To avoid injury during the rocket's ascent, Laika was secured with restraining chains while lying on her belly. Lying on her back would have been the best way to endure the g-force of acceleration, but as Pettit observed, dogs "are not meant to lie on their back. Dogs lie down [on their belly], and a dog isn't used to having a lot of force on their belly either. The geometry of a dog is not something that could take a lot of force. I think it would probably be more unpleasant for a dog [to ride a rocket into space] than it would be for a human being."

The acceleration of the rocket, then, put a lot of force on Laika's belly, making it hard for her to breathe. This is why her respiration increased so dramatically. She was probably taking rapid, shallow breaths, trying to keep air in her lungs as the rocket pressed up beneath her. Once she entered microgravity, the noise and vibration and the tremendous force of acceleration vanished. She would have felt much more at ease, but for her unbearable thirst.

▢

Imagining Laika up there in her spacecraft, confined in that small capsule, makes me feel lonely. This loneliness is my loneliness, of course, but I am not the only one to feel it. Many of the people I spoke with in working on this book were quick to inquire about Laika's loneliness and to express their own loneliness when faced with the thought of her up there alone in space, with the thought of her dying in space. Why do we feel this way when we imagine Laika spinning across the roof of the world, forever interred in that metal can?

Lots of people surrender to animal emotions, dog emotions especially. Can we know how a dog feels when we think of them as feeling like us? In fact, we depend on them to accede to the way we feel, even as we cast our feelings upon them. Knowing this, it is still worth asking: Did Laika get lonely in her spacecraft? Was Laika lonely as she died? Can we say, even, that dogs feel loneliness? Or did Laika feel a sense of purpose in the way that a working dog can? Or perhaps she felt nothing at all. We know that dogs can feel, that they have feelings. Anyone who has ever lived with a dog knows this. We know too that our dogs respond to our feelings. They sense our anger at them, and our anger at each other. Dogs know when people are fighting, when they are in conflict and distress. Dogs pick up these feelings and are then equally in distress. Dogs know when people are happy and at their ease. Dogs then are happy and at ease. A dog knows when another dog is in pain, or when we are in pain, when we are suffering from a wound in the body or a wound in the heart. Our dogs are drawn to our wounds, drawn to us in the center of our grief, drawn to our tears. It follows, then, that dogs might feel lonely.

In a study published by the Association for Psychological Science, social psychologist John Cacioppo of the University of Chicago and his team assert that "loneliness is not a uniquely human phenomenon." Animals of various kinds, especially social mammals, do in fact experience loneliness. Their loneliness, and ours, can be measured by physical responses in the body, namely by "a significant increase in plasma cortisol," a stress hormone. And why do animals feel lonely? Because loneliness is an adaptation, writes Cacioppo; it "represents a generally adaptive predisposition in response to a discrepancy between an animal's preferred and actual social relations that can be found across phylogeny." Loneliness, he writes, "has evolved as a signal to change behavior—very much like hunger, thirst, or physical pain." In manageable doses, then, loneliness triggers a response in the individual to establish or repair social relationships. Social animals need each other, and so loneliness can help keep a group or pair tightly bound, which betters each individual's chances of survival. As an adaptive predisposition, loneliness has been here a long time; it was here on Earth before humans.

When I spoke with astronaut Donald Pettit about his experiences in space, I asked him about loneliness. Pettit is a good man to ask such a question, because he has spent a lot of time in space: 370 days, with thirteen hours of accumulated spacewalk time. In 2003, when the space shuttle *Columbia* broke up as it reentered Earth's atmosphere, killing all seven astronauts on board, Pettit was living and working on the ISS. His mission there was extended to six months, and he had to return to Earth in a Russian Soyuz spacecraft because space shuttle flights had been suspended for the next two and half years. With the space shuttle grounded, I asked him: Did he feel lonely up there in space? Was he not struck by the fear that he might never get home? He is a scientist, a chemical engineer, and I imagine his mind is logically and pragmatically arranged, but even so, everyone gets lonely sometimes.

"There is no sense of loneliness in space at all," Pettit told me. "This is a myth I often get asked about and there's absolutely no loneliness. There's no isolation." He went on to say that on the ISS, you're only 240 miles away from Earth, which is just a bit more than the distance from Dallas to Houston. You can get home in a few hours. "It isn't like you're going to Alpha Centauri or something," the closest star system to ours at 4.3 light-years away, Pettit said. "You are really close to Earth. If you go to some place like Antarctica [where Pettit has

and Literature, "intrinsically alone and irredeemably lost." Everything we do, even our consciousness, is driven by an underlying motivation to avoid loneliness, to avoid isolation. "Forlornness constitutes the very essence of man's existence." And what is so painful about loneliness is that we feel something missing deep in the self, but we do not necessarily know what it is or how to fix it. We do not know what it is that we do not have. Loneliness is the "absence of an awareness of any thing or sensation; it is a meaningless nothingness," Mijuskovic writes. And unfortunately, loneliness is a mainspring of modern civilization. Our ancestors lived in small, intimate, highly communal groups. Civilization, especially Western civilization, has broken and fractured such communities and fed us the lie that strength resides in the individual, alone.

We know that loneliness can be a major health challenge for human beings, that it crushes people and keeps them crushed, and that the lonelier they are, the lonelier they get. Cacioppo has determined that loneliness can contribute to problems like high blood pressure, hardening of the arteries, heart attack, stroke, and even death. You can die from loneliness. Loneliness can cause inflammation in the body; problems with learning and memory; cognitive decline and dementia; a reduction in the efficiency and virus-fighting ability of the body's immune system. Lonely people don't sleep well either, so sleep is not as restorative as it might be. Loss of sleep or interrupted sleep can in turn contribute to more loneliness. And loneliness itself contributes to more loneliness, putting a lonely person in a downward spiral that becomes tougher and tougher to climb out of. The most terrible poverty is loneliness, Mother Teresa taught us. We cannot be prosperous human beings, it turns out, if we are chronically lonely.

Loneliness cannot be avoided, but it can be transcended. To transcend it we must connect with other beings. "As a meaning, the essence of loneliness consists in the overwhelming desire of the yet-unrelated ego to locate, unify, connect, or bind itself in relation to other egos (even animal egos, pets) or objects (e.g., hobbies, amusements)," writes Mijuskovic. The path to transcending loneliness—for human beings and for dogs—is social interaction and acceptance, establishing and maintaining a rich web of relationships, a community, or as Macbeth puts it in Shakespeare's play, "troops of friends."

◌

I was first drawn to Laika's story by imagining what she saw out her window from space, that her view might have offered her relief from the anxiety of confinement, from boredom, and, yes, from loneliness. I imagined the inside of her capsule illuminated by moonshine, and in that brightness, her gaze drawn out onto the bright orb of the moon, and farther still into the blackness of the cosmic deep. She would have seen her reflection too, or a ghost of it, as the dark interior of the capsule was matched by the mostly black of space. I imagined her looking down (or what appears to us as down) onto our cities illuminated on the planet's surface as she passed through the nighttime and then around into day and over the tracks of the Earth's great rivers, long blue spindles running out to the seas, the green of tropical forests, the white matte of winter clouds in the northern hemisphere, circulating storms over the southern oceans, the ocean itself, its impossible blue. I imagined that she saw all that, and as she did, I wondered what she wondered, what she thought about what she saw, and what she felt. I hoped that the view out her window calmed her and gave her a sense of peace into which she died.

In working on this book, I read a great number of books and articles, and I talked with a great number of people who knew Laika's story better than I did. The questions I posed again and again were: Could Laika, in fact, see out her window when she was in space? Was what I had imagined true? Was it Laika, not Gagarin, who was the first living being to see Earth from orbit? Some sources affirmed that Laika could see out her window into space. Others were either not so sure or mostly uninterested in this detail. What did it matter anyway? A few people I spoke with confessed that no one had ever asked them this question, and they simply didn't know. Still other sources led me to a resounding no, Laika would not have been able to see out the window. But I did not want to vacillate in this emptiness of no and yes, and I don't know; I wanted a definitive answer. I wanted proof.

The issue is the fairing on the nose cone of the rocket. When you look at a rocket—United Launch Alliance's dependable Atlas V, for example—you see a long, smooth booster body rising up from the ground, and then about three-quarters of the way to the top, an elongated bulb on the nose of it. This elongated bulb is the payload, the spacecraft or probe or satellite to be deployed in space. Covering it is a protective shell called a fairing. When the rocket enters orbit, the final-stage booster is usually jettisoned along with the fairing, which releases the spacecraft inside. Free from the rocket and its protective shell, the spacecraft can meet its mission goals, whatever those are. The question about Laika's view from her window centers on whether the Soviet team removed the fairing from the satellite when it reached orbit.

Some of the people I spoke with told me that if the goal was to simply get the satellite into orbit it made little sense for the team to take the time (time they didn't have) to install the hardware necessary to remove the fairing. Even if the team had the time, the required hardware and pyrotechnics would have added extra weight to the vehicle. One reason the final-stage booster was not jettisoned from *Sputnik II* was to reduce the rocket's overall weight. Removing the fairing would have added more weight as well as cost and time.

But there is another way of looking at this. "If I were flying that mission, I would have jettisoned the nosecone [fairing]," Gil Moore told me. "Leaving the nosecone on, which covered the satellite, would have interfered with their ability to measure the cosmic ray environment Laika was exposed to in orbit. They would have had better results if the nosecone was removed. And I can't imagine the extra weight of the hardware to remove the nosecone would have made a difference to that R-7 rocket. It had twenty-five engines. Twenty-five!" He continued: "They had to get the nosecone out of the way when they deployed *Sputnik I*, so the hardware to do that had already been developed. And I would think that with *Sputnik II*, they would have wanted to get that nosecone out of the way. But I don't know definitively. I wasn't there, and the Soviets worked in secret. Besides," Moore said, "exposing the window for Laika would have been good preparation for Gagarin, who did have a view of the Earth from orbit."

Amy Nelson, who has written a number of articles about Laika and the Soviet space dogs, agrees. Without offering evidence, she told me she thinks Laika could see out her window.

Where, I wanted to know, was proof?

I discovered an article about *Sputnik II* in *Pravda*, the official newspaper of the Communist Party of the Soviet Union, published ten days after the satellite launched. In that article, I read, "After the last stage of the rocket was established on its orbit, the protective cone was discarded." That sounded clear enough, but during these years *Pravda* was as much propaganda as it was historical record. My research into Laika's story had all along been a slow sifting of misinformation— some of it intentional and some of it misunderstanding—most of which eventually fell away as I continued to read. Still, I could not accept this single source for truth. I needed more.

Then I discovered the work of Anatoly Zak, an authority on Soviet and Russian space history. On his informative website russianspaceweb.com, Zak includes a design drawing of *Sputnik II* with the "payload fairing release mechanism" clearly labeled, along with a short animation of the fairing coming off to expose the satellite. His site also offers an artist's rendering of the satellite in orbit with the Earth visible below. You can see the final-stage booster still attached and the satellite on top of the booster with the fairing removed. The detail is fine enough to identify Laika's capsule. Zak includes the following caption for that image: "Perched on top of a giant rocket, a tiny window could provide a glimpse of the home planet to the first creature ever sent to orbit the Earth." I contacted Zak to discuss this detail and hear more about his sources. Where was this information? Could he help me get to the original source he used? Unfortunately, even after repeated invitations, he declined to speak with me.

Frustrated, I wondered if traveling to the source might give me an answer. In June 2016 I toured the Memorial Museum of Cosmonautics in Moscow. On the first floor near the stuffed space dogs Belka and Strelka is a replica of *Sputnik II*. The outer shell of the satellite is cut away so you can see inside. You can see the cylindrical instrument package on top, the sphere housing the batteries and radio transmitters in the middle, and, nested in the bottom, Laika's capsule with its window. The fairing is perhaps half an inch thick at best. It appeared to me that the fairing was not designed to come off. Of course it is a replica, and some of its design features may have been left out. It is always best to ask. My guide, a space industry professional who worked on the Soviet Union's defunct shuttle program, the Buran, gave me a firm and clear answer: "No," she said in English. "Definitely not. Laika could not see out."

Back in the United States, Cathleen Lewis at the Smithsonian put me in contact with Art Dula, a Houston attorney specializing in intellectual property and space law. Dula was just then in Moscow and soon to meet with a Russian colleague named Evgeny Albats, who for twenty-eight years worked for Zvezda (reorganized as Joint Stock Company in 1994). At Zvezda, Albats worked closely with colleagues at Energia, the company that designed and built Laika's capsule. According to Albats, the fairing on *Sputnik II* was indeed removed in orbit and, he wrote to me, for a time "the dog could see the sky." Excited by this lead, I inquired further. Is the source for this information published? How do I get access to it? Is the source in English, or will I need help with translation? How could my guide in Moscow get this wrong? But Albats could offer no more help.

There must be a published source, I reasoned, that would answer the question definitively. The challenges to finding it had so far kept me adrift in uncertainties: I had little success getting help from people who work in the Russian space industry. I did not have the security clearance or scholarly authority to access Russian archives. I do not speak or read Russian, which, without a translator, made access to these archives useless. I was just a writer trying to tell the story of a dog. The answer to my question remained beyond my reach.

I thought this was the end of it.

Then I raised the matter with Fordham University historian Asif Siddiqi. A Guggenheim fellow who has written a number of seminal works on Soviet space history and exploration, Siddiqi directed me to volume 1 of *Problemy kssmicheskoi biologii* (Problems in space biology), edited by N. M. Sisakian. This Russian-language source (Siddiqi is fluent in Russian) offers a detailed account of the *Sputnik II* mission. It clearly states in a diagram, Siddiqi told me, that the fairing was removed in orbit, exposing Laika's capsule, and so her window, to space.

At last here was my answer, and it posed another question: with the fairing removed and the window exposed to space, what did Laika see?

Before I could approach this question, a final, curious detail came to light. In my correspondence with James Schombert,

spin an object it becomes a kind of gyro itself and highly resistant to changes in attitude. You want the rocket to fly like an arrow shot from a bow, and spinning it around its minor axis, which runs nose to tail, can achieve this. As the rocket gains altitude and enters orbit, it loses energy as its main engines burn out and drop away. Energy dissipation, and imperfections in the load balance of the rocket, causes this spin around the minor axis to become unstable, and the rocket will begin to spin around its major axis, which runs laterally across the center of the rocket body. As the rocket slows down, it begins to tumble end over end. Positioned in the nose cone of the rocket, Laika would have been spinning as if seated in the center of a merry-go-round during the rocket's flight, and when it entered orbit and started to tumble she would have tumbled with it, like riding a Ferris wheel, around and around. The rate of this spin would have been relatively slow, Schombert told me, not above about one rotation per minute, otherwise "centripetal force would have locked Laika to the wall."

It is necessary to take an imaginative leap here, for while we now know that Laika could see out and that the satellite was tumbling in its orbit, we cannot know what she saw, if anything. She was dehydrated and near death from her long delay on the ground, but even in her weakened state, as I see it, the answer to this question is that she saw everything. In the tumbling roll of the satellite, the little window turning over and over on the world, Laika was looking at everything there is, everything there ever was, her eyes taking in starlight that traveled across oceans of time to reach her, light from distant galaxies, from across a billion years, and she was looking down on the living Earth from orbit, even if in a chaotic whirl of the satellite's motion, and she was the first to take in this view. She saw the Earth, the blue marble, fragile, vulnerable, a kind of spaceship itself floating in the black void, impossibly alone. While Ivanovsky claims that Laika was "unaware of what was happening to her and where she actually was," I think she knew. If a dog is anything, it is sensitive and intuitive, and while Laika would not have had our understanding of orbit and space, I think she understood that something big had happened to her. I think she understood—in what way a dog can—that she had crossed the threshold of our world and was now far away. I think she sensed that she had flown into a forbidden realm, far beyond any place anyone had ever been before, a place so foreign, so formidable, so beyond the ken of the ordinary that it would now be nearly impossible for her to come home.

◻

As *Sputnik II* struck a path over the continents, the Soviet Union announced to the world the successful launch and orbit of a living being. As with any news from afar during this time, the initial hours offered inconsistent reports, which were updated as more information came in. Once the West came to understand that the satellite carried a dog, the technological achievements the Soviets were lauding were brushed aside. Never mind that the rocket had lifted over 1,100 pounds into space and so could therefore send a nuclear warhead to America or anywhere else. Never mind that sending an animal into space meant the Soviets would soon be able to send a man into space. Never mind that, as Korolev had claimed, the way to the stars really was open. What was the dog's name, people all over the world wanted to know, and how would it get back home?

Western media considered that the Soviet space dog program had launched dogs into space before and returned them using parachutes. Some had been featured alive and well at press conferences. Some had even given birth to litters of puppies. Perhaps the Soviets would bring the dog home under a parachute or using some other method yet unknown. Perhaps the dog was going to be all right. Other media reports doubted it could come back at all. The spacecraft was in orbit, so how could they get it down? The world stood anxiously by, hoping, even praying, for the dog's safe return.

◻

As Laika made her first and then second orbit, her capsule began to heat up. The data coming into the Soviet ground stations indicated that she was agitated, anxious, moving about, possibly barking. The forced-air cooling system inside the capsule was not keeping pace with the sources of heat: the spent rocket body still attached to the satellite, the electrical systems, the sun, and heat from Laika herself. But there was nothing anyone on the ground could do about it. Some sources indicate that thermal insulation protecting Laika's capsule was damaged, which could have happened when the fairing separated from the satellite. Such damage isn't unheard of, as in the case of the damaged heat tiles on space shuttle *Columbia* in 2003, which led to its destruction on reentry. Both Gazenko and Kotovskaya point to the sun as a major factor in heating *Sputnik II*. Due to its elliptical orbit, the satellite "spent longer in the sun than had been planned," Gazenko said in an interview in *Space Dogs*, "and it began gradually to heat up." In her article, Kotovskaya writes that the "temperature control system inside the capsule was designed so that, while orbiting the Earth, the satellite [would] make it in the shade at times" and cool down. "Unfortunately, the satellite orbit came to be much elongated, elliptical, and most of the time it was in the sunlight."

According to NASA, the temperature range outside the ISS can be as cold as -250 degrees Fahrenheit (on the shady side of Earth), and as hot as 250 degrees Fahrenheit (on the sunny side). The cooling and heating system onboard the space station must be able to manage these extremes. When astronauts work outside the station, performing an "extravehicular activity," also known as a spacewalk, their space suits too must be able to manage this temperature range. Even so, they sense these extreme temperatures through the suit. Temperature control for biological habitats in orbit is an ongoing challenge even in the twenty-first century, so it is not surprising that the Soviets didn't get this quite right on their first try.

Dogs do not manage heat well. Normal body temperature range for dogs is 101 to 102.5 degrees Fahrenheit. Anything above that will cause an increase in heart rate and respiration (excessive panting), restlessness complicated by lethargy, and possibly vomiting and diarrhea. A body temperature of 106 degrees is often the line between life and death. If a dog's body temperature rises to that level, it will likely die, and die soon if it cannot cool down. Dogs sweat only from a gland in the bottom of their feet, which is why some dogs will stand in cool water, or in their water dish, after vigorous exercise in warm weather. They also pant, gassing off heat and circulating cooling air through their mouth and nose and over their tongue. But even better than sweating from their feet and panting is to find protection and relief from the heat in cool water, in shade, or indoors.

John Smith, a veterinarian in Texas who allowed me to observe him performing both a spay and a neuter surgery, told me that heat exhaustion can become critical in dogs very quickly. If you are going to save a dog in heat distress, Smith said, you have to bring their temperature down rapidly, usually by immersing the dog in an ice bath. The dog is then very weak, vulnerable, and highly sensitive to temperature. Once cooled, the dog must be removed from the ice bath and stabilized, or else its temperature will continue to fall and the dog will die going in the other direction. If the dog can't cool down, either on its own or with help, it will become increasingly lethargic and disoriented, fall into a coma-like state, and die. It doesn't take long, perhaps ten or fifteen minutes, Smith told me, depending on how hot it is and how fit the dog is, and so how efficient it is at cooling itself. Humidity increases the heat challenge and the process can move along more rapidly. I asked Smith what dying from heat feels like for a dog. He looked at me for a moment and said, "Pain."

◻

After three times around the Earth, the temperature in Laika's capsule had climbed to 104 degrees Fahrenheit and possibly as high as 109 degrees. Shortly after that, the data from Laika's sensors stopped coming in. There was no sign of respiration or pulse, no sign of anything except the satellite itself, streaking across the sky, crossing continents in minutes, speeding faster than anything humans had ever built. The Soviets celebrated the Revolution and the state that manufactured it, a miracle really, the satellite, the flight of the first living being in space, the new religion of science and technology that had so elevated the Soviet Union, at least in those days and months, to the status of the greatest nation on Earth.

I have imagined Laika in her capsule speeding around the planet in the mostly dark, the unseen immensity of the cosmos surrounding her. She would have started panting as the temperature rose, a little at first, then heavily as her temperature rose with the temperature of the capsule, her heart rate rising too, her eyes closing with the cooling action of her panting as she struggled to manage the heat. The more she panted, the more she raised the humidity inside the capsule and in turn the temperature, overwhelming the cooling and air regeneration system—a greenhouse effect. She could not move much, seated there in that pod, wires pulled out from beneath her skin, whirling through space. Seated there, her eyes began to close in her growing lethargy, and an agitation too rising in her with the temperature, because as a young dog she still possessed a great deal of energy, and yet she could do nothing about that, nothing to quiet the heat, to end the threat she felt coming on. Her thirst would have been immense. She could not move away from the heat or seek shelter from it, the shade of a tree perhaps or the sanctuary of her home. Perhaps she wanted for some of that space dog food, that gelatinous glob set before her at the feeding tin because it had given her a little relief. But I imagine that her appetite left her as the temperature climbed, and if she had had more food in those last moments, she would not have touched it. All she could do was sit there, her eyes falling heavy and closing, not from the action of her panting now but because she was drifting in and out of a melancholic delirium that settled in, her eyes filling with the flashes of light that come in the darkness in space, and her head no longer in her control, floating on its stem, her paws stretched out before her to the front of the capsule just a few inches from the shores of the universe. Then Laika fell into a coma from which she would never awake. And that was all.

◻

When the team came to understand that Laika was dead only a few hours into her flight, they continued to report on the success of the launch and the satellite. The Soviets had already been criticized for their space dog program, so they remained silent about Laika's demise. When I asked Sergei Khrushchev why the team had kept Laika's death secret, he replied, "It was not kept secret. Because we were not publicizing it does not mean it was kept secret. There was not a way to bring the dog back, so it was understood that the dog would be sacrificed for science. No one knew, really, that the capsule had overheated at the time. It was of no interest to anybody to announce that the dog died, but rather to celebrate the launch." There was no conspiracy here, Khrushchev was telling me, so much as a continued focus on the achievements of *Sputnik II*. And those achievements were real. Laika was always going to die, so did it matter that she died sooner or later, before or after? The satellite was in orbit, and no nation on Earth had achieved such a towering technological feat except the Soviet Union, and they had now done it twice.

Still, the Soviets were practiced at neglecting to report or even record embarrassing failures. It was policy to demonstrate Soviet power and achievement to the world, as it was policy to do the same in the United States. What benefit was there in offering the long trail of trials and mistakes? "It's mostly lost now in the enormity of those achievements," Dubbs told me, "but [Laika's death] was a failure of technology, or rather of Soviet philosophy. Laika was sacrificed as much for political expedience as for advancing the space program, and the Soviets did not want to advertise their failures."

The day after the launch, the media reported that the dog—who was still without a name in most countries outside the Soviet Union—was doing fine and might yet be recovered. The next day, November 5, the satellite dominated the news again, with articles speculating about the dog's return, articles expressing outrage at sending a dog at all, and articles supporting the choice as a necessary step in human spaceflight. Turkina writes in *Soviet Space Dogs* that Radio Moscow issued this statement to the greater Soviet Union: "Although we are filled with sympathy and sorrow for little Laika, at the same time we cannot divert our attention from the enormous significance of her sacrifice for scientific research." The announcement does not indicate knowledge of Laika's death but of her eventual death. It is a call for sympathy and sorrow in preparing the public for that inevitability, at least in the Soviet Union.

Protests broke out in the West. In London members of the National Canine Defense League met with the first secretary of the Soviet Embassy to protest and called for a worldwide daily minute of silence until Laika was returned home. In New York City a picket line formed in front of the United Nations offices. In a telegram to the Soviet Embassy in Washington, DC, an animal adoption group characterized *Sputnik II* as an atrocity. In *Sputnik: The Shock of the Century*, Dickson reports that the American Society for the Prevention of Cruelty to Animals deplored using a dog to test a technology that could not "possibly advance human health and welfare." A *New York Times* editorial characterized the dog in space as the "shaggiest, lonesomest, saddest dog in all history." On November 6 an Australian newspaper reported that the dog in the satellite was called Laika.

Soviet scientists were still reporting telemetry signals coming in from *Sputnik II* as late as November 7, which was in fact true. Laika was dead, but onboard instruments were sending back information about radiation levels and cosmic rays in orbit. On November 10 *Sputnik II*'s batteries went flat and all telemetry transmissions ceased. Radio Moscow affirmed this on November 11, and then on November 12 the Soviets announced to the world that Laika was dead.

For the next several decades various Soviet publications offered conflicting accounts of Laika's death. The Soviets acknowledged that Laika could not have been brought back, and so they had euthanized her quickly, painlessly, humanely. Some sources say that Laika had been fed a poison in her food. Others say a poisonous gas was released into her capsule. Some report she was injected automatically with a poisonous serum. Still other reports had Laika dying of asphyxiation when the oxygen supply in her capsule ran out. These stories faded with time as the world turned to other affairs. Then in 1993, at work on *Animals in Space*, co-author Colin Burgess met with Oleg Gazenko in Vienna at the Association of Space Explorers congress. In that conversation, Gazenko confirmed that Laika had died "soon after launch" of heat exhaustion.

If the Soviets had developed the technology to return Laika to Earth and made the decision to bring her back when her capsule started heating up, it would have done little good. She would have died anyway. There was not enough time to get her to the ground and release her from the capsule to cool her down. The team had discussed how to deal with a problem like this, how to euthanize Laika by some mechanism they controlled from the ground so that she didn't suffer. "We had wanted an option to kill Laika," Seryapin said in *Space Dogs*. "Laika would be put out of her misery within a few seconds. Some say it was planned that way, but I don't know. It didn't happen, and Laika died a slow and painful death that lasted about an hour and a half to two hours."

I think Laika's death lasted much longer than two hours. I think she started dying the moment they put her in the capsule. After the first day inside the capsule, while still on the ground, she began to suffer from dehydration despite the team's efforts, despite their love and care for her. By the time she entered orbit she had been inside the capsule for three days with only the water in her space dog food and the little water she received from Yazdovsky and Alexander Dmitrievich through the breathing hole. By the time the temperature began to rise inside her capsule in orbit, her suffering must have been at an end, for she would have mostly been gone already. For Laika to survive in space for seven days was an impossibility. She had no chance.

The story of Laika's death, when and how she died, did not come to light until 2002. Dimitri Malashenkov, who had worked on the *Sputnik II* project, gave a paper entitled "Some Unknown Pages of the Living Organism's First Orbital Flight" at the World Space Conference in Houston, Texas. With this paper, Malashenkov explained to the world for the first time that Laika had died not after seven days in orbit but after several hours. "After ground simulation of the flight conditions the conclusion has been made," he writes, "that Layka should be lost from an overheating on 3–4 circuit of flight," meaning between the third and fourth orbit. Written in what appears to be Malashenkov's English, not in translation, he goes on to remark that "it was practically impossible to create a reliable system of a temperature control in such small term."

☐

When you travel to a place that is not your own, it is best to have a guide. You need help from a guide who knows where you are going, who has been there before, who can point out its dangers and pleasures and subtleties. On our journey into space, Laika was that guide for us. She was our scout, a star dog, a cyborg, a highly trained cosmonaut, the first cosmonaut, the first space traveler, an explorer in her own right. We followed her into Earth orbit and from there found our courage and journeyed to the moon. After the hollow years of World War II and on into the Cold War night terrors of global nuclear devastation, we all needed something to believe in, something to sustain our broken spirits. A trained soldier of the Cold War, a war dog, Laika emerged from that hostility to forge a new season of cooperation in space between the USSR, the US, and other nations, whose governments had so polarized the world, so cultivated a climate of fear, that we were living in the shadow of our mutual destruction. Had those rockets all been missiles, we would have no science in space, no space exploration; we would have only war. But a dog—even a war dog—doesn't believe in war. A dog believes only in the task at hand. In space, Laika flew over all our troubles, all our pettiness, and opened a window on our world and on the cosmos, and through that window we could all see that the grand design was so much grander, more mysterious, more vast and empty and dark and filled with light than we had imagined. We came to understand that only in combining our resources—our science and technology, our political wills, our economies, our cultures and our art—could we explore that mystery. Laika was not a lab animal, not the subject of experimentation, not a victim of human ambition. She was an extension of the men and women who trained her, an exten-

◌

Epilogue

The Earth is the cradle of humanity, but one cannot live in the cradle forever.

KONSTANTIN TSIOLKOVSKY
personal letter, 1911

If we can get to Mars, we can go anywhere.

STEPHEN PETRANEK
How We'll Live on Mars, 2015

In his lecture "Mankind in the Universe," given before the German and Austrian Physical Societies in Salzburg in 1969, American theoretical physicist and mathematician Freeman Dyson marked the beginning of the Space Age as June 5, 1927, when "nine men meeting in a restaurant in Breslau [now Wrocław, Poland] founded *Verein für Raumschiffahrt*" (VFR), the Association for Space Navigation. The VFR was a private upstart without government funding that for six years "carried through the basic engineering development of liquid-fueled rockets." Hitler shut it down, but as Dyson sees it, the VFR stands as the "first romantic age in the history of spaceflight."

Then in 1958, another grassroots organization, this time in the United States, formed around a theoretical physicist named Ted Taylor, who wanted to build a spacecraft powered by nuclear explosions. Dyson, one of the key members of the team, stated in his 1969 lecture that the organization had the spirit of the VFR in mind, and they began with three principles:

1. The conventional von Braun approach to space travel using chemical rockets would soon run into a dead end, since manned flights going farther than the moon would become absurdly expensive.

2. The key to interplanetary flight must be to use nuclear fuel, which carries in each kilogram a million times as much energy as chemical fuel.

3. A small group of people with daring and imagination could design a nuclear spaceship that would be both cheaper and enormously more capable than the best chemical rocket.

The group called itself Project Orion. "We felt from the beginning," Dyson said, "that space travel must become cheap before it can have a liberating influence on human affairs." In Project Orion, Dyson saw a better use of the world's stockpiles of nuclear weapons: "We have for the first time imagined a way to use the huge stockpiles of our bombs for better purpose than for murdering people. My purpose, and my belief, is that the bombs which killed and maimed at Hiroshima and Nagasaki shall one day open the skies to man."

Ted Taylor, whose story is the subject of John McPhee's 1973 book *The Curve of Binding Energy*, called the H-bomb the worst invention ever but also the most interesting. He was attracted to extremes in physics, and nuclear explosions are about as extreme as it gets. Space exploration and science go side by side with the bomb and its delivery system, the missile, with one key difference: while highly competitive, space exploration and science also encourage cooperation among nations, as opposed to conflict, to achieve goals. Turner saw an opportunity to transform humankind's most destructive power into one of its most constructive. Dyson agreed wholeheartedly, suggesting that Project Orion was "not only a scientific instrument but an imperative for the future of the world," writes McPhee. While one danger of possessing the awesome power of nuclear weapons was using them against each other, Dyson also identified another danger fundamental to the human species. "He saw the human race running out of frontiers," McPhee writes, "and he considered frontiers essential to the human psyche, for without them pressures would build that would implode upon the race and destroy it." Space—that final frontier—Dyson was saying, might save us from ourselves.

Taylor's vision was a spaceship containing two thousand nuclear bombs that dispensed through a hole one at a time and exploded beneath the ship, propelling it upward. Building the bombs was the easy part. In order to learn how to deploy them, Project Orion consulted with Coca-Cola to understand their coin-operated Coke dispensing machines. The bottom of the spacecraft had to be constructed of something that could take multiple nuclear blasts. Taylor considered steel, copper, aluminum, and wood. The team resolved that some kind of fiberglass might work best. The bombs would be of increasingly higher yield, pushing the ship upward out of the Earth's atmosphere, until the fiftieth bomb, which, "at twenty kilotons," writes McPhee, "would be of the force that destroyed Nagasaki." Such immense power (unachievable using liquid- or solid-fueled rockets) could lift a payload of a thousand tons into orbit and then beyond. The team imagined a spaceship shaped like a bullet about the size of a ten-story building with all the amenities of a cruise ship inside: state rooms, exercise and recreation rooms, a restaurant with an observation deck, a level devoted to gardens and food animals. The ship would launch from Jackass Flats, Nevada, or possibly from a barge at sea, and it would be, Taylor told McPhee, "the most sensational thing anyone ever saw." With Orion, the team determined, human beings would land on Mars by 1965 and travel to Saturn by 1970.

In 1961 Taylor traveled to Huntsville, Alabama, and presented his idea to Werner von Braun, who became its powerful proponent. A believer in the necessity of a voyage to Mars, von Braun had written a book on the subject more than ten years earlier, *Das Marsprojekt*, which was later translated into English as *The Mars Project*. It is, writes Stephen Petranek in *How We'll Live on Mars*, "a work of extraordinary foresight and sheer engineering genius."

The US Air Force backed Project Orion but had to justify it as a military endeavor. A spacecraft powered by nuclear explosions could also resist nuclear explosions fired at it, and so could safely rain down bombs on the enemy without fear of counterattack. McPhee writes that the Strategic Air Command's general Thomas Power said, "Whoever builds Orion will control the Earth!" Faced with this irony—a project founded on using nuclear bombs to elevate and advance human life instead of killing people, backed by a military that wanted to use that same technology to kill people—Taylor was convinced that they were close to getting it done. McPhee quotes Taylor: "Just a few little twists of events and everything we were trying to do with Orion would have come through." Once up and running, both Taylor and Dyson were convinced, Orion would throw open the door to the mysteries of the universe, and that new knowledge would safeguard life on Earth.

But Project Orion was anathema to many scientists and engineers, then and now, especially when it comes to safeguarding life on Earth. Detonating nuclear bombs in the Earth's atmosphere produces radioactive fallout, radiation that rains back down onto Earth. By the early 1960s scientists were measuring strontium-90 in cow's milk, a radioactive isotope prevalent in fallout from the routine testing of nuclear bombs, primarily by the US and the USSR. When ingested, strontium-90 finds its way into bones and can cause bone cancer, cancers of the surrounding tissues, and leukemia. "It was well known by the early 1960s, the dangers of massive amounts of radiation being dumped into the atmosphere," astronomer James Schombert told me. "How this was ignored by Orion designers reminds me of the way tobacco companies ignored the lung cancer statistics." President Kennedy signed the Partial Test Ban Treaty in 1963, banning nuclear explosions underwater, in the atmosphere, and in space, which killed Project Orion but did not kill the dream of Mars.

○

For the remainder of the 1960s, the US and the USSR were locked in a race to put men on the moon. When the US won that race with the Apollo program, the USSR turned to research projects on crewed space stations in Earth orbit. After the final moon landing in 1972, the US joined the USSR in that effort. To move supplies and crews back and forth between Earth and orbit, the Soviets employed the same basic rocket technology they had used to put up the *Sputnik* satellites, and the US went to work on the space shuttle program. (Concurrent with the US space shuttle program, the Soviets were working on a similar shuttle orbiter, the Buran. After one successful unmanned flight in 1988, the program was canceled with the dissolution of the USSR.) While the space shuttle has often been called a marvel of technology, it was also limited and limiting. It proved to be "fragile, expensive, and dangerous," writes Guy Gugliotta in "Space: The Next Generation," his 2007 story for *National Geographic.* "And since it cannot fly beyond low-Earth orbit, it has transformed spaceflight into a series of high-tech cruises to nowhere." After the Apollo moon missions, von Braun argued fervently for a mission to Mars, and with Ted Taylor's ideas in mind he thought he could do it using nuclear-powered spacecraft. From the time of Laika's flight into orbit, it took just over a decade to put men on the moon. During that time of great technological advancement and creativity driven by Cold War competition, if the US had pushed past the moon instead of investing in the space shuttle, many space industry experts argue, we would have a working outpost on Mars today. Instead of Mars, President Nixon backed the space shuttle program, because in addition to its science mission it could be used to deploy and repair spy satellites. This choice, writes Petranek, "threw the US space program into a long, slow de-

makes a compelling case. We must go to Mars to learn about
Mars, which will in turn teach us about the Earth. We must
go for the challenge, because we are a species that thrives on
challenge. For most of human history, war has been that challenge and a prime motivation for advances in medicine and
technology. But we have before us an opportunity to replace
war with the challenge of cosmic exploration. "The time is
past for human societies to use war as a driving stress for technological progress," states the declaration. We must go for the
future of humanity, not for ourselves but for generations yet
unborn. A permanent settlement on Mars will revitalize our
belief in and our valuing of human life. Elon Musk agrees.
"Musk is in love with the idea that humans should become a
spacefaring society," writes Stephen Petranek in *How We'll Live
on Mars*. SpaceX's mission in these early decades of the twenty-first century sounds a lot like Ted Taylor's Project Orion
from the 1950s: make space travel more affordable and go to
Mars. In late 2016 Musk rolled out his detailed plan to colonize
Mars, shuttling people and equipment in SpaceX spacecraft.
China, India, Russia, Japan, the European Space Agency, private industry, and NASA have all been stirred to think and
move in this direction. China has even appointed an ambassador to Mars, NBA star Yao Ming. Mars will be, Musk tells us,
the greatest adventure in the history of humankind.

With the end of the space shuttle program in 2011, the
United States lost its capability to put astronauts into orbit.
We have relied on Russia to get that done, buying rides for
about $75 million a seat. Both private industry and NASA have
been working on regaining crew launch capability. SpaceX is

almost there, and NASA is developing the Space Launch System, designed to propel an Apollo-like crew capsule capable of long-duration deep-space travel, with a Mars mission targeted for some time in the 2030s. It isn't quite the spacecraft Ted Taylor imagined, or von Braun before and after him, but it might do the job. NASA calls their new crew capsule Orion.

◌

The famed British scientist James Lovelock told the *Guardian* in 2008 that climate change is unstoppable and all there is left to do is enjoy our lives, because if we're lucky we've got about twenty years before "it hits the fan." Frank Fenner, a professor of microbiology at Australia National University, has written that because of climate change and related effects, humans will be extinct by 2100. In *Learning to Die in the Anthropocene*, Roy Scranton writes that due to climate change, human "civilization is already dead. The sooner we confront our situation and realize that there is nothing we can do to save ourselves, the sooner we can get down to the difficult task of adapting, with mortal humility, to our new reality."

Beyond climate change, there are more dangers to the long-term survival of our species. Physicist Stephen Hawking had said we have about a thousand years left on Earth, and he later revised that number to one hundred years. Hawking cites climate change and overpopulation as major threats, as well as the development of powerful technologies, especially artificial intelligence (AI), that may incite catastrophic wars, either using AI against each other for control of the planet or wars against AI. Stephen Petranek has developed a list of some ten ways our world might end suddenly, among them a planet-killing asteroid, which would strike Earth with the force of many Hiroshimas and raise a dust cloud blotting out the sun that would kill all the plants on Earth that life depends on. Like the dinosaurs, we would not escape such an event, and such an event is a statistical inevitability. It's just something that happens from time to time, and as of yet we have no defense against it. As Carl Sagan has written, human beings have but two choices: "spaceflight or extinction."

Elon Musk too "is keenly aware that Earth will not be habitable forever," writes Petranek in *How We'll Live on Mars*. "Musk seems frustrated by our denial about what we are doing to our habitat [the Earth], and is ever cognizant of a simple fact: humans will become extinct if we do not reach beyond Earth." And reaching beyond Earth means colonizing Mars. It is the only option now, maybe the only option ever. The runaway greenhouse effect on Venus makes it far too hot (some 900 degrees Fahrenheit on the surface), and the moons of Saturn and Jupiter are too far away, at least for our initial effort. At about a six-month journey one-way, Mars is not much farther off than Lisbon was from Rio de Janeiro in the sixteenth century. While conditions on Mars make it challenging, it is achievable. Establishing a colony there will be, writes Petranek, "nothing less than an insurance policy for humanity." Or as Musk has put it, a backup for the biosphere.

◻

The first two decades of the twenty-first century have been a boon to space science and exploration. We—the human animal—have landed the rover, *Curiosity*, on Mars to explore the Gale Crater and environs (2012); landed the *Philea* probe on the comet 67P/Churyumov-Gerasimenko (2014); discovered that liquid water flows on Mars, making it ever more likely that life was or still is present there (2015); flew the *New Horizon* probe past Pluto to explore that astonishing and active world at the limit of our solar system (2015); inserted the *Juno* probe into orbit around Jupiter to find out more about that planet's formation and origin, and thus the formation and origin of Earth (2016); detected gravitational waves pulsing outward from the collision of two massive black holes at an immense distance from the Earth, waves that Einstein's theory of general relativity predicted a hundred years ago (2016); discovered that Earth has a second moon, a little asteroid called 2016 HO3, that our planet caught about a hundred years ago and will likely hang on to for a couple hundred more (2016). We accomplished all this using robotic technology. No one had to travel to the moon, and no one had to travel to Mars.

When I began to explore Laika's story, I felt strongly that human beings should not go to Mars. We should stay put and work on things at home, because the fate of humanity is inseparable from the fate of the Earth. We should use our capital, ingenuity, intelligence, and cooperative spirit to build a more sustainable civilization here on Earth, one that will stabilize and then reduce human population growth, repair our decimated ecosystems, and encourage biological diversity. Humans will do better, because the Earth will do better. When I began to explore Laika's story, I wanted to make a plea for an expansion of robotic exploration of our solar system and an end to crewed missions into space. Robots are cheaper and hardier than humans, and they don't require training or life-support systems. Instead of the expense and resources required to send humans to Mars or anywhere else, we should send robots and robots alone, robots that are, as Carl Sagan has written of his beloved Voyager spacecraft, "intelligent being[s]—part robot, part human," that "[extend] the human senses to far-off worlds." Without humans in space, we might also reduce the need for biological research in space. We would not need to learn how to survive long-term in microgravity or how to shield ourselves from deadly radiation in space or on Mars, and we would no longer need to send animals of any kind into space. If biological research did continue in space, it would be directed away from how humans might live out there and toward how humans and all Earth's creatures can live better here. That research would help us take care of our planet, not help us leave it. When I began to explore Laika's story, I wanted to say that space exploration is already at its limit, that Mars is a fantasy and will always remain one. The dream of Mars, I wanted to say, is no better than a wish for immortality, a wish for what we can never have. I wanted to say that what we really need as a species is to stay home, on Earth. What we really need is to be at home on Earth.

But I feel differently now.

Freeman Dyson is right—he must be right—that the human animal needs, always, a new frontier to push against. We cannot prosper without the exercise of our spirit of discovery and exploration. We need to explore to remain whole—physically, emotionally, psychologically. Maybe even spiritually. Going to Mars is not for everyone, but everyone will be struck with awe and amazement when we do. Nothing good will come of a suppression of human desire. It will manifest itself, either in a positive way that unites us and inspires and improves life on Earth, or in a destructive way that continues to polarize nations, fuel bloody competition for resources, and ensure the end of us all. Restraint is not the way. The way is expression, release, liberty, Mars.

When I spoke to astronaut Donald Pettit, I asked him what he thought about a crewed mission to Mars, about establishing a permanent colony there. Is Mars even possible? "It's inevitable," he said. "Human beings will become scattered throughout the solar system." At the present rate of development and planning, he said, it will take a couple hundred years but we can easily accelerate this process if we have the financial and political will.

Having come this far, having gone that far out, how can we turn our back on human space exploration, on the adventure that is Mars, when such adventures have the power to free us from our myopic and self-serving perceptions? How can we turn our back when such an adventure may bring us to understand again that we are in the universe together, and it is our togetherness—our empathy for all living creatures—that sustains us? Instead of squabbling and warring on Earth, all of humanity might join in the project to colonize Mars. How can we turn our back on Mars when, as Hubert Planel writes in *Space and Life*, "exploring faraway hospitable planets directly has always been Man's dream"?

☐

John Karas, vice president of business development at Lockheed Martin, believes it will require a cooperative effort to colonize Mars. "Space is a matter of national pride," he said. "Nations want to be in space because it is the technological high ground. But I don't see a Space Race today so much as a space cooperation, or collaboration. Likely Mars will be a joint venture among a consortium of nations. We can't do it alone. To go to Mars, we will need to collaborate with spacefaring nations. This will be a human endeavor, not an American or a Russian endeavor."

After the divisive US presidential election of 2016—which is more a thermometer for the health of our country, even for the health of our species, than a cause for its illness—it's clear that our government is failing, at least in the short term, to unite us as a people. But perhaps Mars can do it. Perhaps Mars can do even more. Perhaps Mars can unite the US with the rest of the world, especially with our longtime partners in space exploration, Russia. For the past seven decades or so, cooperation in space has mostly transcended whatever political harangue is in play. Even with tensions building between the US and Russia over Syria, charges of Russian interference in the 2016 US presidential election, Russia's annexation of Crimea, US sanctions against Russia, Edward Snowden, you name it, Russian president Vladimir Putin made this statement in 2016: "We attach great importance that despite whatever difficulties we face on Earth, people in space work shoulder to shoulder, hand in hand, to help each other and fulfill tasks that are essential not just for our countries but for all of humanity." Maybe we need a Mars mission to bring the world together in common cause. Mars could be this century's greatest creative act, the renewal of the world's science and technology, political and economic relationships, art and artistic form. Mars can be the physical expression of everything that is good about us.

"Yes!" Karas told me when I asked him what he thought of these ideas. "I love these ideas. Without cooperation, we're going nowhere. And the technology required to go somewhere, to go to Mars for example, will benefit everybody on Earth." Consider the many benefits that came from the Apollo missions to the moon, he said. That technology required "us to build everything smaller. Those missions kick-started a microelectronics revolution. We need a Mars mission to push technology. If we don't have such a mission, the work, the innovation and discovery doesn't happen. To get to Mars and to live there, we will have to learn how to get water from a rocky, barren desert planet, where there is, as we are now discovering, a lot of water in the form of ice. We will have to dramatically improve our solar technology, and we will have to find ways to protect astronauts from radiation. Now do you think we could use these same technologies here on Earth?" Not to mention, Karas said, "the cultural dynamics, the shift in the way we think about the Earth that will come from such an endeavor. A journey to Mars will give us a perspective of our own planet that would otherwise be impossible. This paradigm shift will come in like a lion, a shift as big as our coming to understand that the Earth is not the center of the universe. Space is a realm of peace and sharing. Space exploration is for

two little moons, Phobos, at about fourteen miles across, and Deimos, nearly eight miles across. Mars's gravity is about 37 percent of Earth's, which means you can jump about three times as far. Mars has the highest known mountain in the solar system, a shield volcano called Olympus Mons, which rises 15.5 miles above the Martian surface (Everest is just over 5.5 miles). The mountain covers an area of approximately 120,000 square miles, or just shy of the size of the entire state of New Mexico.

⬚

Sputnik II, and *Sputnik I* before it, changed everything. What we were before we ventured into space we can never be again. Spacefaring made us into something else, but something no less essential to ourselves. So fully have we occupied this new self as to shut off the possibility of return. There is no going back. There is only the going on. And if things go along as Konstantin Tsiolkovsky imagined, along with Robert Goddard, Sergei Korolev, Ted Taylor and Freeman Dyson, Werner von Braun, Elon Musk, and others, we *are* going on. We are going to Mars, and we are going to colonize it. We are going to establish permanent habitation on the Red Planet, and future generations of humans will be born and live out their lives there. Future generations will be able to immigrate from Earth to Mars. You can see this truth emerging in the books we're writing, in the movies we're making, in the science and planning in the space industry, and in the excitement of people all over the world.

Mars is a planet without animals, and when we go to Mars I do not think we will go alone. We have always taken animals with us into space, and in fact we sent them into space ahead of us. When we go to Mars, we will fill our ships with biological experiments. Fruit flies will certainly go, mice, microbes (those we choose, and those that hitch a ride with our equipment and our bodies), plants of various kinds, and a store of seeds, some for science and some for crops. Maybe we'll take a few species of fish, molds and yeasts (we could not live on Mars without beer), and maybe bees. Wouldn't we want some bees? We'll take animals for food—chickens, lambs, maybe goats or hogs—a veritable farm of animals to seed the Red Planet with their kind. The biblical story of Noah's Ark is not history, and it is not myth. It's prophecy, and the only thing that may stop us from colonizing Mars is both the reason we must go and the reason we may not make it: our own extinction.

Animals for science and food will not be enough. There is no hope of joy—on Earth or on Mars—except in companionship, with other people and with other animals. Human evolution and development are inseparable from the evolution and development of animals, especially the dog. I think we will take our dogs to Mars when we go. For how could we live without them? Why would we want to? On our first journeys into space we followed the tracks of dogs, and when we go to Mars we will make those tracks with them. When the people of Earth gaze up at that small red point in the morning sky that is Mars, and the first Mars colonists stand with their dogs looking back at the beauty of the Earth, it will be with gratitude to Laika, who showed us all the way.

BIBLIOGRAPHY

Abadzis, Nick. *Laika*. New York: First Second, 2007.

Andrews, James T., and Asif A. Siddiqi, editors. *Into the Cosmos: Space Exploration and Soviet Culture*. Pittsburgh: University of Pittsburgh Press, 2011.

Asashima, Makoto, et al., editors. *Fundamentals of Space Biology*. Tokyo: Japan Scientific Societies Press, 1990.

Bergwin, Clyde R., and William T. Coleman. *Animal Astronauts: They Opened the Way to the Stars*. Englewood Cliffs, NJ: Prentice-Hall, 1963.

Berry, Wendell. "The Body and the Earth." In *The Art of the Commonplace: The Agrarian Essays*, 99. New York: Counterpoint, 2003.

Brzezinski, Matthew. *Red Moon Rising: Sputnik and the Hidden Rivalries That Ignited the Space Race*. New York: Times Books, 2007.

Burgess, Colin, and Chris Dubbs. *Animals in Space: From Research Rockets to the Space Shuttle*. Chichester: Praxis, 2007.

Chernov, V. N., and V. I. Yakovlev. "Research on the Flight of a Living Creature in an Artificial Earth Satellite." *ARS Journal Supplement* 29, no. 10 (1959): 736–42.

Chertok, Boris. *Rockets and People*, edited by Asif Siddiqi. 4 vols. Washington, DC: NASA History Series, 2012.

Conn, P. Michael, and James V. Parker. *The Animal Research War*. New York: Palgrave Macmillan, 2008.

Conquest, Robert. *Kolyma: The Arctic Death Camps*. New York: Viking, 1979.

Costlow, Jane, and Amy Nelson. *Other Animals: Beyond the Human in Russian Culture and History*. Pittsburgh: University of Pittsburgh Press, 2010.

D'Antonio, Michael. *A Ball, a Dog, and a Monkey: 1957—The Space Race Begins*. New York: Simon & Schuster, 2007.

Dickson, Paul. *Sputnik: The Shock of the Century*. New York: Walker, 2001.

Dyson, Freeman. "Mankind in the Universe." Lecture at Salzburg, September 29, 1969. *Physikalische Blätter* 26 (1970): 7–14.

Gillispie, Charles Coulston. *The Montgolfier Brothers and the Invention of Aviation, 1783–1784*. Princeton: Princeton University Press, 1983.

Gugliotta, Guy. "Space: The Next Generation." *National Geographic Magazine*. October 2007.

Harford, James. *Korolev: How One Man Masterminded the Soviet Drive to Beat America to the Moon*. New York: John Wiley & Sons, 1997.

Helvajian, Henry, and Siegfried W. Janson. *Small Satellites: Past, Present, and Future*. El Segundo, CA: Aerospace Press, 2008.

Henderson, Caspar. *The Book of Barely Imagined Beings: A 21st Century Bestiary*. Chicago: University of Chicago Press, 2013.

Horowitz, Alexandra. *Inside of a Dog: What Dogs See, Smell, and Know*. New York: Scribner, 2009.

In the Shadow of the Moon. Directed by David Sington. Discovery Films, 2007.

Ivanov, Aleksei (Oleg Ivanovsky). *Pervye Stupeni: Zapiski Inzhenera (The First Steps: An Engineer's Notes)*. Moscow: Molodaya Gvardiya, 1970. Portions translated for Kurt Caswell by Natalia V. Maximova.

Jacchia, L. G. "The Descent of Satellite 1957 Beta One." Smithsonian Astrophysical Observatory Special Report no. 15. July 1958.

Khrushchev, Sergei N. *Nikita Khrushchev and the Creation of a Superpower*. Translated by Shirley Benson. University Park: Pennsylvania State University Press, 2000.

King-Hele, D. G., and Mrs. D. M. C. Walker. "The Last Minutes of Satellite 1957 Beta (Sputnik 2)." *Nature*, August 16, 1958, 426–27.

Kotovskaya, Adil Ravgatovna. "Why Were Flying Dogs Needed for Rockets and Satellites to Launch Yuri Gagarin?" Received via email from the author. Publisher unknown.

Kurnosova, L. V., editor. *Artificial Earth Satellites*. 2 vols. New York: Plenum Press, 1960.

Launius, Roger D., et al., editors. *Reconsidering Sputnik: Forty Years since the Soviet Satellite*. Amsterdam: Overseas Publishers, 2000.

Linenger, Jerry M. *Off the Planet: Surviving Five Perilous Months aboard the Space Station Mir.* New York: McGraw-Hill, 2000.

Malashenkov, Dimitri C. "Some Unknown Pages of the Living Organisms' First Orbital Flight." IAF Abstracts, Astrophysics Data System, 2002. 288.

Mayberry, Jim, editor. *The New Mexico Museum of Space History Curation Paper No. 8.* 2012.

McPhee, John. *The Curve of Binding Energy.* New York: Farrar, Straus & Giroux, 1973.

Mondry, Henrietta. *Political Animals: Representing Dogs in Modern Russian Culture.* Leiden: Brill Rodopi, 2015.

Murakami, Haruki. *Sputnik Sweetheart.* Translated by Philip Gabriel. New York: Vintage International, 2001.

Oberg, James E. *Red Star in Orbit.* New York: Random House, 1981.

Parker, James V. *Animal Minds, Animals Souls, Animal Rights.* Lanham, MD: University Press of America, 2010.

Petranek, Stephen L. *How We'll Live on Mars.* New York: TED Books, 2015.

Planel, Hubert. *Space and Life: An Introduction to Space Biology and Medicine.* Boca Raton, FL: CRC Press, 2004.

Rhea, John, editor. *Roads to Space: An Oral History of the Soviet Space Program.* Translated by Peter Berlin. New York: McGraw-Hill, 1995.

Scranton, Roy. *Learning to Die in the Anthropocene: Reflections on the End of Civilization.* San Francisco: City Lights Books, 2015.

Shelton, William. *Soviet Space Exploration: The First Decade.* New York: Washington Square Press, 1968.

Siddiqi, Asif A. *Challenge to Apollo: The Soviet Union and the Space Race, 1945–1974.* Washington, DC: National Aeronautics and Space Administration, 2000.

———. "Iskusstvennyy sputnik zemli." *Spaceflight* 49 (November 2007).

Soviet Writings on Earth Satellites and Space Travel. London: MacGibbon & Kee, 1959.

Space Dogs. Directed by Ivan Mazeppa. BBC Four, 2009.

Turkina, Olesya. *Soviet Space Dogs.* Translated by Inna Cannon and Lisa Wasserman. London: Fuel Design, 2014.

Verne, Jules. *Around the Moon*. Hertfordshire: Wordsworth Editions, 2011.

———. *From the Earth to the Moon*. Hertfordshire: Wordsworth Editions, 2011.

Wang, Xiaoming, and Richard H. Tedford. *Dogs: Their Fossil Relatives and Evolutionary History*. New York: Columbia University Press, 2008.

Witt, Peter N., et al. "Spider Web-Building in Outer Space: Evaluation of Records from the Skylab Spider Experiment." *Journal of Arachnology* 4 (1977): 115–24.

Wolfe, Tom. *The Right Stuff*. New York: Bantam Books, 1979.

ACKNOWLEDGMENTS

This book's heart is rooted in the imagination of one of its editors at Trinity University Press, the poet Barbara Ras. I am deeply grateful for Barbara, for her life and her life's work, for her belief in me as a writer, for calling me out of the void to take a ride on this star. She has made my life better, and I will remain in gratitude until the end.

My gratitude to everyone too at Trinity for championing my work, especially director Tom Payton and my editor Stefanie Mortis, whose talent and job it is to make books better— she certainly made this book better. Also at Trinity, thanks to Burgin Streetman, Sarah Nawrocki, Lee Ann Sparks, and Bridget McGregor.

Chris Dubbs, Natalia V. Maximova (Natasha), and R. Gilbert Moore (Gil) each offered their time and resources beyond the measure of ordinary generosity. Chris is one of the original champions and scholars of the space dogs of Soviet Russia, and his help and guidance have been invaluable. To aid in my research, Natasha translated a key resource from Russian and offered me a deep and critical reading of the early manuscript. When I traveled to Moscow, she arranged visits at museums and sites important to Soviet and Russian space history, as well as other sites near Moscow that I was interested in. She took me into the countryside to visit friends at their dacha, where I hunted mushrooms in the forest and experienced the noble tradition of the Russian banya. I could not have traveled in Russia with such ease and confidence without her help. Gil's technical mastery, enthusiasm, and careful feedback helped steer my revision process. I was delighted to discover in our discussions that many of the people I was writing about were Gil's friends and colleagues, and many of the events in this book are central to his life. Thanks too to Gil and his wife, Phyllis, for welcoming me into their home for good conversation and good cheer.

I am grateful to a good number of people for their time and expertise: Tom Macarrone in the Physics Department at Texas Tech University; Hwihyun Kim and Judit Gyorgyey Ries at the University of Texas; Sergei Khrushchev; Adil Ravgatovna Kotovskaya; Amy Nelson; veterinarian John Smith and the staff at ARK Hospital in Lubbock; Jeff Thomas at the Texas Tech Health Sciences Center; Erin Collopy and Anthony Qualin in the Texas Tech Department of Classical and Modern Languages and Literatures; Asif Siddiqi at Fordham University, who helped answer one of the central questions of this book; John McGlone, Alexandra Protopopova, and Glenna Pirner in the Texas Tech Department of Animal and Food Sciences; former chairman and CEO of Lockheed Martin, Norm Augustine, a man of depth and vision; Patricia Alvarez, Patricia McCarroll, John Karas, David Welch, and Andrea Greenan at Lockheed Martin; Jim Mayberry and Sue Taylor at the New Mexico Museum of Space History; Fred Wyman, Sarah Knoll, and everyone at United Launch Alliance; Catherine Tsairides at Wyle; the man whose team designed, built, and installed the wiring harness on *Atlas 5 OA-6* and with whom I shared a bus seat to the launch site at Cape Canaveral; my uncle Dick and aunt Janet Cantwell in Ocala, Florida, who took me in and talked my ears off; Brandi Dean and astronaut Donald Pettit at NASA; Cathleen Lewis at the Smithsonian National Air and Space Museum; James Schombert at the University of Oregon; Art Dula, Elena Perepelkina, and Evgeny Albats, who contributed vital information that helped sharpen this story's clarity and focus; Richard Porter at the Texas Tech International Cultural Center; the Texas Tech Office of Research for a generous grant that helped me with my journey to Russia; Steve Fritz; P. Michael Conn at the Texas Tech Health Sciences Center and Alice Young at Texas Tech for their help in understanding the place of animals in scientific research; John Lane; Michael San Francisco and Aliza Wong in the Texas Tech Honors College, where I teach, who opened doors, bent spoons with their minds, and carved out space, time, and money for me in support of this book; Kathryn Miles, a writer of grace and sensitivity who knows a hawk from a handsaw, a situation from a story; my parents, who gave me a room with a woodstove to work through the winter; Karen Clark, a talented editor and gracious soul, for her support, counsel, and love, from beginning to end; Scott Dewing, my brother along the Way—*dokkodo*—for traveling with me to Russia and then down the slope of Eastern Europe; and Kona, the German shepherd who watched over me as I wrote this book in the many places I wrote it—Texas, Florida, Saskatchewan, Idaho, Oregon, and dozens of little camps between here and there where I set up in my Alaskan camper and made words into sentences into paragraphs into pages, and into this book. Thank you.

Kurt Caswell is a writer and professor of creative writing and literature in the Honors College at Texas Tech University, where he teaches intensive field courses in writing and leadership in the mountains and on rivers in the West. His books include *Getting to Grey Owl: Journeys on Four Continents* and *In the Sun's House: My Year Teaching on the Navajo Reservation* (both published by Trinity University Press), and *An Inside Passage*, which won the 2008 River Teeth Literary Nonfiction Book Prize. He is the co-editor of *To Everything on Earth: New Writing on Fate, Community, and Nature.* His essays have appeared in *ISLE, Isotope, Matter, Ninth Letter, Orion, River Teeth, McSweeney's, Terrain,* and the *American Literary Review.* He lives in Lubbock, Texas.